Optical Glass

American Institute of Physics
Translation Series

Optical Glass

T. S. Izumitani
Hoya Corporation
Tokyo, Japan

New York 1986

Originally published as *Kogaku Garasu*
© 1984 Kyoritsu Shuppan, Ltd.

This book was translated from the Japanese by the Berkeley Scientific
Translation Service.

Library of Congress Cataloging in Publication Data

Izumitani, Tetsurō, 1924–
    Optical glass.

    Translation of: Kōgaku garasu.
    Includes bibliographies and index.
    1. Glass, Optical. I. Title.
QC375.I9813 1986   666'.156   86-22236
ISBN 0-88318-506-7

Foreword

When I moved to the Osaka Industrial and Technical Experimental Institute from Kyoto University at the age of 24, I decided to follow in the footsteps of Dr. Susumu Takamatsu and to make Japanese optical glass the best in the world, surpassing Schott of Germany. In the more than 30 years that have passed since then, I have devoted my life to the research and development of optical glass. My move to Hoya Glass (K.K.) during this period stemmed from a desire to realize the continuous melting of optical glasses. I think that my expectations have been more or less fulfilled.

My basic attitude toward research and development is to "build technology on the basis of science." My publications, "The Vitreous State," "Dispersion in Optical Glasses," "Continuous Melting of Optical Glasses," "Polishing of Optical Glasses," "Physical Properties of Laser Glasses," etc., arose from my inclination to seek the underlying essence of things. The themes that I have pursued are the determination of the bases of technology and the building of new glasses on the basis of physical properties.

This little work of mine is not a collection of data, nor is it an all-inclusive textbook. It is a book that describes what appears to me to be the essence of optical glasses and might well be titled "My View of Optical Glasses." I am well aware that this little book of mine has its shortcomings. Nevertheless, I have dared to publish it because I hope that this small work might become a milestone in the history of optical glasses. It is one small stone placed in the history of optical glasses. If people who follow in the future find a source for development in its pages, I will be more than happy.

I have received the cooperation of many friends and co-workers in this project. I am deeply grateful to Mr. Ryohei Terai and Mr. Yoshiro Moriya, who conducted basic research on optical glasses; to Mr. Hiroshi Ogawa, who cooperated with me in the development of continuous melting; to Mr. Shin'ichi Hirota, who conducted research on dispersion; to Mr. Kenji Nakagawa, who cooperated with me in research on phase separation; to Mr. Shoichi Harada, who conducted research on polishing; to Mr. Hisayoshi Toratani, who cooperated with me in research on lasers; to Mr. Yoshiyuki Asahara, who always offered his fullest cooperation; and to Dr. Akira Ikushima and Dr. Hiroto Kuroda at the Institute of Solid State Physics of Tokyo University, and Dr. Ryumo Onaka at Tsukuba University, who were always obliging with their kind guidance. I am also grateful to the government Industrial Research Institute, Osaka, and Hoya Glass (K.K.) for providing facilities for this research and development. I deeply appreciate my good fortune in having so many friends and co-workers, and such excellent work facilities.

T. Izumitani
Christmas morning, 1983

Preface to the English Translation

I am pleased and happy to have my life's work published in this translation of my book, *Optical Glass*. I started my research on optical glass composition by investigating the basic science of glass formation, refractive index, and dispersion. Later my work included research on the continuous melting of optical glass. My findings in this area revolutionized the optical-glass manufacturing process that had previously been based on the claypot melting process developed by O. Schott in Germany. I have also been interested in the glass-polishing process. In this area, I found that the glass is polished by the formation of a hydrated layer. This layer is removed by scratching the surface with polishing grains embedded in pitch.

Since publishing the original book, I have investigated reflection spectra in the vacuum ultraviolet (VUV) and extreme ultraviolet (EUV) regions. Before my investigation, VUV and EUV were, so to speak, in a black box because we had not located the absorptions of nonbridging and bridging oxygen. In addition, we did not know the difference between the bridging of absorption wavelengths in silicate, borate, and phosphate glasses, or between oxide and fluoride glasses. I was able to measure the reflection spectra in the VUV and EUV using synchrotron orbital radiation in the range of 10 to 40 eV. Because I was lucky enough to measure these reflection spectra, I now understand the relationship between dispersion and absorption of optical glass.

Recently, I began working on a glass molding process. This new method completely changes the lens manufacturing process, which previously consisted of grinding and polishing. With my process, I can obtain precise spherical or aspherical surfaces without deformation. To achieve these surface conditions, I press glass blanks in a mold for more than 20 seconds until a homogeneous temperature distribution (at a temperature where the viscosity is in the range of $10^{7.6}$ to 10^{13} poise) occurs throughout the lens.

In other current work, I have found that the high cross section of stimulated emission of phosphate laser glass comes from the asymmetry of the crystalline field surrounding Nd^{3+} ions. The asymmetry is due to the chain structure of phosphate glass.

My optical glass investigations are very exciting to me. Although my work on optical glass is small, I am very happy that I have made some contribution to the science of optical glass, and, therefore, to the scientists and engineers involved in this field throughout the world.

I am indebted to Dr. J. L. Emmett, Lawrence Livermore National Laboratory (LLNL), for having my book translated into English. It would have been impossible to publish my book in English without his enduring friendship. I also appreciate Dr. S. E. Stokowski's (LLNL) kind reading and correcting of the many drafts of this manuscript.

T. Izumitani
Fremont, California

Acknowledgments

The American Institute of Physics gratefully acknowledges the assistance of the Technical Information Department of Lawrence Livermore National Laboratory in the development of this project; special thanks are due to Leonard Fisher, Peter Murphy, Elaine Price and Peter Link.

Contents

CHAPTER 1

What is Glass?

We know that three states of matter exist: gas, liquid, and solid. What kind of state is the vitreous state? The drinking glass that you see sitting on the table is transparent like water, but it does not flow like water. Even if you are told that "glass is a supercooled liquid," it does not slosh around like water does so it does not seem to be a liquid at all. Even if it is stated that "glass is a supercooled liquid and is in a metastable state," the drinking glass on the table does not change with time, no matter how many days or years may pass, and it looks as if it must be in a state of equilibrium. What is glass?

1.1. The Vitreous State Viewed Thermodynamically

1.1.1. The Vitreous State

When changes in specific volume and thermal capacity are investigated for a glass melt that has been slowly cooled, the results obtained are as shown in Figs. 1.1 and 1.2. As the temperature drops, the specific volume decreases. When the melting point T_f is reached, an ordinary liquid releases its fusion heat and changes into a crystal. Afterward, the volume decreases in accordance with the expansion coefficient of the crystal. However, a substance that solidifies into a vitreous state does not crystallize at the melting point, but continues to be cooled in a liquid state. In other words, it is supercooled. Accordingly, this state is metastable; thermodynamically, it is not a stable state. A supercooled liquid, however, has a fixed volume at a constant temperature regardless of

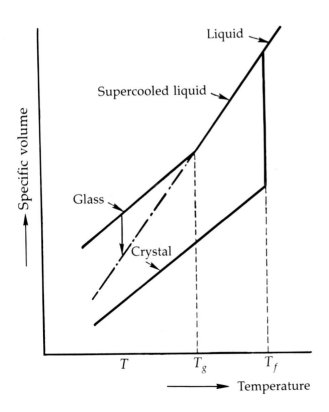

Figure 1.1. Model of the relationship between specific volume and temperature.

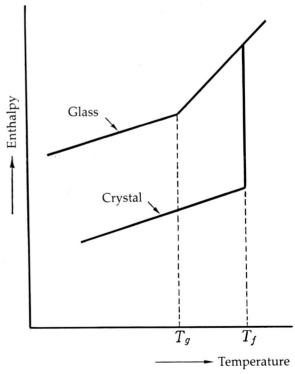

Figure 1.2. Model of the relationship between enthalpy and temperature.

1

time. As the temperature drops even further, though, the change in volume becomes more gradual, as shown in Fig. 1.1. In the end, the volume decreases with almost the same expansion coefficient as in a crystal.

The temperature T_g at which a shift is seen from the expansion coefficient of a liquid to that of a solid is called the transition point. Expressed as a coefficient of viscosity, this corresponds to 10^{13} poise. A supercooled liquid below this temperature is called a vitreous state. Above this temperature, the liquid is simply in a supercooled state.

When a glass is maintained at a constant temperature (e.g., T) below the transition point, the volume decreases with time until it reaches a certain equilibrium value. This phenomenon is called stabilization. If the maintenance temperature is in the vicinity of the transition point, a longer time is required for equilibrium to be reached as the temperature is lowered. At ordinary room temperatures, the time required is close to infinity. The equilibrium values at various temperatures are on an extension of the volume-temperature curve of the supercooled liquid. When a glass is cooled at a constant rate, the relaxation time of the glass structure is increased. As a result, the structure cannot follow changes in temperature and is frozen at a certain fixed temperature. At ordinary temperatures, therefore, glass structure shows no change over time.

In other words, glass can be defined as a substance in which a supercooled state is maintained at temperatures below the melting point so that no devitrification (crystallization) occurs, and in which the structure is frozen in the vicinity of the transition point.

As shown in Fig. 1.1, a glass has a greater specific volume than a corresponding crystal at ordinary temperatures. Similarly, a glass also shows higher values of enthalpy than a crystal (Fig. 1.2).

When various thermodynamic quantities of $Li_2O \cdot 2B_2O_3$ glass and B_2O_3 glass are compared with those of the corresponding crystals, using the relation

$$\Delta F = \Delta H - T\Delta S \quad ,$$

we can say that glass is in a thermodynamic state in which the free energy, entropy, or enthalpy is somewhat higher than in the corresponding crystal (see Table 1.1).[1]

1.1.2. Glass Structure

A glass is a supercooled liquid. How does its structure differ from that of a crystal? As shown in Fig. 1.3 (for example), quartz crystal forms a network structure in which SiO_4^{4-} tetrahedrons are regularly linked in hexagonal form. In quartz

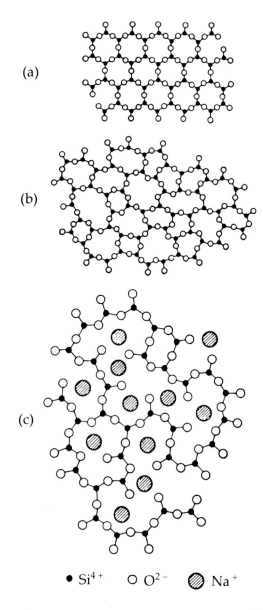

(a)

(b)

(c)

● Si^{4+} ○ O^{2-} ◍ Na^+

Figure 1.3. Network structure of SiO_4^{4+} tetrahedrons: (a) quartz crystal, (b) quartz glass, (c) sodium silicate glass (Ref. 5).

Table 1.1. Thermodynamic quantities of B_2O_3 type glasses.

	ΔH_f (kcal/mol)	ΔH_{25} (kcal/mol)	$\Delta S(S_{vit} - S_{cry})_{25}$ (cal/deg/mol)	ΔF_{25} (kcal/mol)
$Li_2O \cdot 2B_2O_3$	28.77	11.16	4.62	9.78
B_2O_3	5.5	4.35	5.71	2.65

glass, on the other hand, the interatomic bonding angle is not constant, nor does the network show a fixed polygonal form; rather, an irregular network structure is formed. It has been stated that when modifier ions (Na^+, Ca^{2+}, etc.) are introduced, the network structure is cut and alkali ions are positioned in the holes of the network (Zacharisen[2] and Warren,[3] network structure theory). In other words, it seems that in glass the irregular atomic arrangement of a liquid is frozen. In contrast to the irregular network structure theory, there is also a theory that proposes the existence of microscopic regular structures in glass (Porai-Koshits[4]).

1.2. Crystallization and Glass Forming Substances

1.2.1. Crystallization

To explain why a glass can be supercooled, let us first consider the process of crystal deposition from a melt. Below the melting point, the chemical potential of a solid is certainly lower than that of a liquid (Fig. 1.4). However, when solid microcrystals are deposited, new surfaces are formed at those points. Unless the difference in chemical potential between the liquid and the solid is greater than this surface energy, the microcrystals cannot exist in a stable state. Furthermore, for the microcrystals to grow, material must be supplied from the liquid. In ordinary substances, the rate of diffusion of the liquid is high so that the substance immediately solidifies into a crystal when cooled below the melting point. In the case of glasses, however, the viscosity is high; accordingly, even if the substance is supercooled, crystal nuclei do not grow, and the substance remains in a supercooled state. Let us deal with this point a bit more quantitatively.

Let us consider the transition from phase A to phase B in Fig. 1.5. For each molecule of phases A and B, the chemical potential is assumed to be ϕ_A, ϕ_B, respectively, and the surface energy is assumed to be σ. If phase B consists of spheres of radius r, and the numbers of molecules of A and B

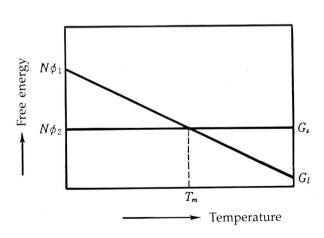

Figure 1.4. Relationship between free energy and temperature for a liquid and solid (Ref. 6).

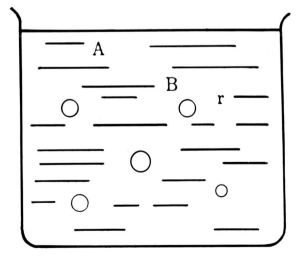

Figure 1.5. Initial nuclei in the liquid phase.

are N_A and N_B, respectively, then the thermodynamic potential Φ of the system as a whole is given by the following formula:

$$\Phi = N_A \phi_A + N_B \phi_B + 4\pi r^2 \sigma \quad . \qquad (1.1)$$

In a state of equilibrium, $\Delta\Phi = 0$, and $N_A + N_B = $ a constant. Accordingly,

$$\phi_B - \phi_A + 4\pi\sigma\left(\frac{dr^2}{dN_B}\right) = 0 \quad .$$

If the volume of one molecule of phase B is v_B, then

$$N_B = \frac{4\pi r^3}{3v_B} \quad ,$$

$$(1.2)$$

$$\therefore \phi_B - \phi_A + \frac{2\sigma}{r} v_B = 0 \quad .$$

Now, if the melting point (or liquidus temperature) is T_l, then the difference in chemical potential when the liquid is supercooled to temperature T is as follows:

$$\Delta F = \phi_B - \phi_A = \Delta H - T\Delta S \quad .$$

If the heat of fusion of one molecule is λ, then at the melting point it is

$$\Delta H(\equiv -\lambda) = T_l \Delta S \quad ,$$

$$\therefore \Delta S = \frac{-\lambda}{T_l} \quad .$$

Accordingly,

$$\phi_B - \phi_A = (T_l - T)\Delta S = \frac{(T_l - T)\lambda}{T_l} \quad .$$

Or from Eq. (1.2),

$$T_l - T = \frac{-T_l}{\lambda}(\phi_B - \phi_A) = \frac{2\sigma v_B T_l}{\lambda r} \quad . \qquad (1.3)$$

This equation illustrates the relationship between the degree of supercooling and the size of nuclei present at the time. As described below, substances that become glass have a low liquidus temperature and a small degree of supercooling. That is to say, such substances can be character-

ized as substances in which the difference in free energy between liquid and solid is small, and in which it is difficult for crystal growth to take place unless the nucleus size r is relatively large. Next, let us consider the process and rate of crystal growth.

The difference in thermodynamic potential between a case in which nuclei are present and a case in which no nuclei are present (i.e., the work required for nucleation) is as follows:

$$\Delta\Phi = \Phi - \Phi_0$$

$$= \left(N_A\phi_A + N_B\phi_B + 4\pi r^2 \sigma\right) - \phi_A\left(N_A + N_B\right)$$

$$= -\left(\phi_A - \phi_B\right)N_B + 4\pi r^2 \sigma \quad .$$

Because $N_B = 4\pi r^3/3v_B$,

$$\Delta\Phi = -\frac{\phi_A - \phi_B}{v_B}\frac{4\pi r^3}{3} + 4\pi r^2 \sigma \quad .$$

Under conditions of equilibrium, from Eq. (1.2),

$$\Delta\Phi = 4\pi\sigma\left(-\frac{2}{3}\frac{r^3}{r^*} + r^2\right) \quad .$$

The term $\Delta\Phi$ is maximum when $r = r^*$; accordingly,

$$\Delta\Phi_{max} = \frac{4}{3}\pi r^{*2} \sigma \quad . \qquad (1.4)$$

The above is the activation energy for nucleation; only microcrystals that cross the $\Delta\Phi_{max}$ potential barrier can grow. Microcrystals of size r^* are called nuclei (Fig. 1.6).

Then, from Boltzmann's distribution law, the number of phase B nuclei is determined as follows:

$$N_{r*} = \text{const} \cdot \exp\left[\left(-\Delta\Phi_{max}\right)/(kT)\right]$$

$$= \text{const} \cdot \exp\left[\left(-4\pi r^{*2}\sigma\right)/(3kT)\right] \quad .$$

However, since the formation of nuclei of size r^* is regulated by the activation energy for diffusion (ΔU), the number of crystal nuclei N_c formed per unit time can be expressed as follows:

$$N_c = \text{const} \cdot \exp\left(-\Delta U/kT\right)$$

$$\cdot \exp\left[\left(-4\pi r^{*2}\sigma\right)/(3kT)\right] \quad .$$

From Eq. (1.3),

$$r^* = \frac{2\sigma v_B T_l}{\lambda(T_l - T)} \quad,$$

$$\therefore N_c =$$

$$\text{const} \cdot \exp\left\{\frac{-1}{kT}\left[\Delta U + \frac{4\pi\sigma}{3}\left(\frac{2\sigma v_B T_l}{\lambda(T_l - T)}\right)^2\right]\right\} \quad.$$

$$(1.5)$$

The variable N_c increases with an increase in $(T_l - T)$, i.e., with an increase in the degree of supercooling. On the other hand, the rate of diffusion decreases with a drop in temperature. Therefore, N_c reaches a maximum value at a certain temperature. Similarly, the crystal growth rate (formation and growth of two-dimensional nuclei) is as follows:

$$\phi_B - \phi_A + \frac{\sigma}{r}(v_B)^{2/3} = 0 \quad, \tag{1.6}$$

$$\Delta\Phi_{\max} = \pi r^* \sigma \quad. \tag{1.7}$$

Accordingly, the growth rate I is given by the following equation:

$$I = \text{const} \cdot \exp\left[-\frac{1}{kT}\left(\Delta U + \frac{\pi\sigma^2 T_l v_B^{2/3}}{\lambda'(T_l - T)}\right)\right] \quad. \tag{1.8}$$

The relationship between temperature and number and growth rate of nuclei has been experimentally determined by Tamman.[7] This relationship is shown in Fig. 1.7. These facts allow us to state that glass is a substance that, because of a low liquidus temperature and high viscosity, shows no crystal growth and remains in a liquid state even when supercooled. We must mention that, in addition to the small difference in free energy between liquid and solid, another factor that makes a supercooled liquid or glass possible is its high viscosity. In other words, a glass-forming material must be a highly viscous substance, that is to say, a substance with strong interatomic bonds.

1.2.2. Glass Forming Substances

Many researchers have discussed the problem of what kinds of substances can form a glass. Early on, the following conditions were proposed as conditions for glass formation in an empirical rule established by Zacharisen[2]:

- An oxygen ion coordination number of 3 or 4 for the cations forming the network.
- An irregular three-dimensional network formed by these triangular or tetrahedral structural units with the corner oxygen ions as common points.

Glass forming substances have also been considered in terms of chemical bonding. According to Dietzel,[8] such substances can be classified into glass forming oxides (glass formers), intermediate oxides (intermediates), and modifier oxides (modifiers) according to z/a^2, where z is the electric charge and a is the interionic distance between the

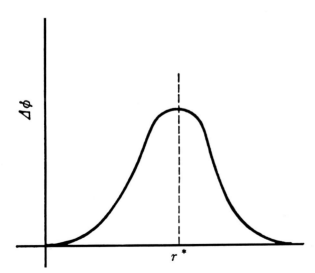

Figure 1.6. Relationship between the difference in thermodynamic potential and radius of nuclei (embryo).

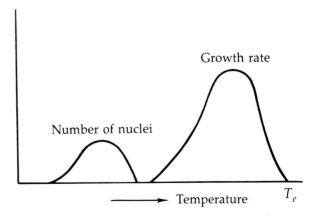

Figure 1.7. Relationship between the number and growth rate and the temperature of crystal nuclei.

cation and oxygen ions: P^{5+}, B^{3+}, Si^{4+}, Ge^{4+}, etc., are glass forming oxides; Al^{3+}, Be^{2+}, etc., are intermediate oxides; and Mg^{2+}, Zn^{2+}, Ca^{2+}, Pb^{2+}, Na^+, etc., are modifier oxides. Values of z/a^2 are shown in Table 1.2. In general, cations with a high valence and a small ionic radius form covalent bonds with the surrounding oxygen ions. A similar classification can be obtained according to single bond strength[9] (a value obtained by dividing the heat of dissociation by the coordination number) or electronegativity.[10] The electronegativity of O^{2-} is 3.5; the greater the difference between the electronegativity of the oxygen ion and the negativity of the cation, the more ionic the bond. These values are also shown in Table 1.2. The fact that interatomic bonding is strong here is probably responsible for the high viscosity, which constitutes one of the conditions for vitrification.

The above-mentioned concepts on the conditions for vitrification are based primarily on geometrical, chemical-bonding considerations. However, vitrification can also be considered from the standpoint of dynamics. That is to say, if "glass" is interpreted to mean a substance that has a low liquidus temperature and a high viscosity, and in which rapid cooling cannot cause crystal growth, then a broader general view may be taken regarding vitrification. For example, we must admit that in the vitrification of $ZnCl_2$ (Ref. 11) (which is not generally thought of as a glass forming oxide), the

vitrification that occurs at the central part of ternary systems as shown in Fig. 1.8 (Ref. 12), and the recent appearance of amorphous metals are phenomena that cannot be interpreted in terms of the tetrahedron theory. A more satisfactory interpretation appears to be that "glass is a liquid that has solidified without ever having time to crystallize."

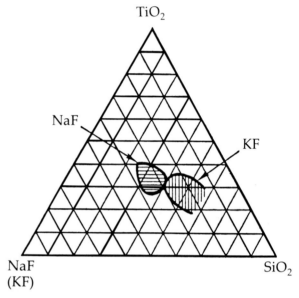

Figure 1.8.   Vitrification regions in NaF– and KF–TiO$_2$–SiO$_2$ ternary systems (Ref. 12).

Table 1.2.   Glass forming substances and physical properties that act as measures.

	Ion	Ionic bonding of oxide (%)	Coordination number	$2z/a^2$	Electro-negativity	Single bond strength (kcal)
Glass forming oxides	P^{5+}	39	4	4.3	2.1	88 to 111
	B^{3+}	42	3	3.22	2.0	119
	Si^{4+}	50	4	3.14	1.8	106
	Ge^{4+}	55	4	2.65	1.8	108
Intermediate oxides	Al^{3+}	60	(4)	1.69	1.5	53 to 67
	Be^{4+}	60	4	1.51	1.5	63
Modifier oxides	Mg^{2+}	70	6	0.95	1.2	37
	Zn^{2+}	—	4	0.91	—	36
	Ca^{2+}	75	8	0.69	1.0	32
	Sr^{2+}	75	8	0.58	1.0	32
	Pb^{2+}	—	6	0.53	—	39
	Ba^{2+}	80	8	0.51	0.9	33
	Li^+	75	4	0.45	1.0	36
	Na^+	80	6	0.35	0.9	20
	K^+	81	9	0.27	0.8	13
	Cs^+	82	12	0.22	0.7	10

1.3. Phase Separation

1.3.1. Introduction

In addition to the crystallization described above, phase separation is another phase transition phenomenon that occurs when a glass has been cooled into a supercooled liquid state. Phase separation (liquid-liquid separation) does not occur in all glasses. Whereas crystallization is a liquid-solid phase-transition phenomenon, phase separation may be thought of as a phase transition phenomenon consisting of liquid-liquid separation. Phase separation was discovered from the formation of liquid drops in the heat treatment processes of crystallized glasses[13,14] and is now being studied.

1.3.2. Thermodynamics of Phase Separation

1.3.2.1. Phase Separation in Binary Systems

In a binary system consisting of components a and b, the internal energy of the system is given by the following equation (where x is the concentration of component a, $1 - x$ is the concentration of component b, and s is the coordination number around the respective atoms):

$$U = \frac{Nsx^2}{2}E_{AA} + \frac{Ns(1-x)^2}{2}E_{BB} + Nsx(1-x)E_{AB}$$

$$= \frac{Ns}{2}\left[xE_{AA} + (1-x)E_{BB}\right.$$

$$\left. + 2x(1-x)\left(E_{AB} - \frac{E_{AA} + E_{BB}}{2}\right)\right].$$

Here, N is the number of a and b atoms, and E_{AA}, E_{BB}, and E_{AB} indicate the respective energies of A-A, B-B, and A-B bonds.

The increase in the entropy of mixing, resulting from mixing the above two components, is as follows:

$$\Delta S (= k \ln W) = -Nk\left[x \ln x + (1-x)\ln(1-x)\right].$$

Accordingly, if the increase in the free energy of the system is expressed as

$$\Delta G = \frac{Ns}{2}\left[xE_{AA} + (1-x)E_{BB}\right.$$

$$\left. + 2x(1-x)\left(E_{AB} - \frac{E_{AA} + E_{BB}}{2}\right)\right]$$

$$+ NkT\left[x \ln x + (1-x)\ln(1-x)\right], \quad (1.9)$$

$$E_{AB} - \frac{E_{AA} + E_{BB}}{2} = w, \quad (1.10)$$

then w acts as a measure of the tendency toward phase separation: if $w > 0$, then the two components tend to separate (see Fig. 1.9b); if $w < 0$, then the two components tend toward homogeneity (see Fig. 1.9a). In the former case, the separation of a solution of composition x into x_1 and x_2 results in a lower free energy of the system; accordingly, the solution is separated into two solutions indicated by x_1 and x_2. This phenomenon is liquid-liquid separation of the solution. When two solutions are homogenized, heat is generated; in the case of separation, heat is absorbed.

The strength of the tendency toward liquid-liquid separation is more or less determined by the value of w. The value of w is determined by the difference in bond energy between network-forming ions (Si^{4+}, B^{3+}, etc.) and modifier ions (Na^+, Ca^{2+}, etc.) with respect to the O^{2-} ions. In practical terms, the tendency toward liquid-liquid separation can be ascertained by an approximation of bond energy using cationic field strength. Specifically, when the difference in cationic field strength is large, the solution tends toward homogenization; when the difference between the two field strengths is small, the solution tends to separate.

1.3.2.2. Phase Separation in Ternary Systems

In ternary systems,[15] the behavior of the third component with respect to phase separation is not simple. For example, SiO_2–Li_2O systems, SiO_2–Al_2O_3 systems, and Al_2O_3–Li_2O systems all individually show a tendency toward separation; as Al_2O_3 is progressively added to a SiO_2–Li_2O system, however, phase separation is suppressed. Assuming that the phase separation temperature is T_c, the relationship in Eq. (1.11) is obtained when a third component c is added.[16] [Note that R is the gas constant.]

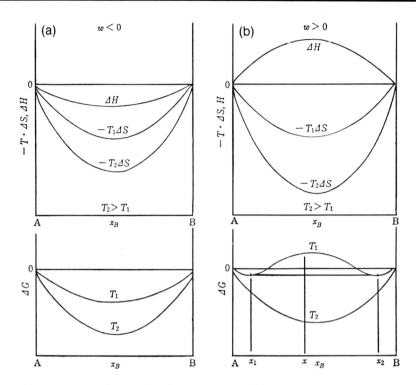

Figure 1.9. Change in free energy with temperature in binary systems: (a) tendency toward homogenization and (b) tendency toward separation.

$$\frac{\delta T_c}{\delta x_c} =$$

$$- \frac{1}{2R} \frac{\left(\alpha_{AB} - \alpha_{BC} + \alpha_{AC}\right)\left(\alpha_{AB} + \alpha_{BC} - \alpha_{AC}\right)}{\alpha_{AB}} .$$

$$(1.11)$$

Here, α_{AB} can be expressed as follows:

$$\alpha_{AB} = E_{AB} - \frac{E_{AA} + E_{BB}}{2} ,$$

α_{AC} and α_{BC} are also similarly defined.

As shown in Fig. 1.10, the tendency toward phase separation of A and B is increased by component c when $\alpha_{AB} < |\alpha_{AC} - \alpha_{BC}|$, whereas the tendency toward such phase separation is decreased when $\alpha_{AB} > |\alpha_{AC} - \alpha_{BC}|$.

1.3.3. Dynamics of Phase Separation (Growth Mechanisms of Liquid-Liquid Separation)

Two regions with different phase separation properties have been discovered to exist within

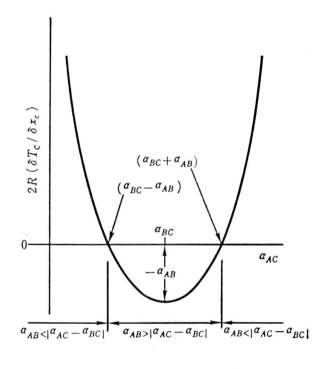

Figure 1.10. Relationship between $\partial T_c / \partial x_c$ and α_{AC} (Ref. 15).

the phase separation region of a binary system. One of these regions occurs where $\partial^2 G/\partial x^2 > 0$ (shown in Fig. 1.11). Here, the free energy increases with a very small fluctuation in composition so that work is required to separate the system into two liquid phases. Liquid-liquid separation proceeds by a mechanism in which nuclei are first produced and then grow (nucleation-growth mechanism). In the region where $\partial^2 G/\partial x^2 < 0$, on the other hand, the free energy decreases with a very small fluctuation in composition so that no work is required to separate the system into two liquid phases. Accordingly, liquid-liquid separation is controlled only by diffusion (separation by spinodal decomposition). Table 1.3 shows the differences between phase separation by the nucleation-growth mechanism and phase separation by spinodal decomposition.

1.3.3.1. Phase Separation in the Metastable Region

Phase separation in this region proceeds by a mechanism involving the formation of phase-separation nuclei and the growth of liquid drops. Such separation can be treated in exactly the same way as the formation and growth of crystal nuclei. Specifically, the rate of formation I_c of liquid-drop nuclei can be expressed as follows:

$$I_c = \text{const} \cdot \exp\left\{ -\frac{1}{kT}\left[\Delta U + \frac{16}{3}\frac{\pi\sigma^3}{\Delta S_v^2 (\Delta T_m)^2}\right]\right\} .$$

(1.12)

Here, ΔU is the activation energy of diffusion, ΔT_m is the amount of supercooling from the mixing temperature T_m, and ΔS_v is the change in entropy

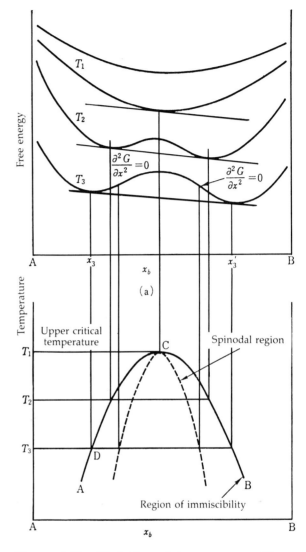

Figure 1.11. Variation of free energy with respect to temperature in each of the two phase separation regions.

Table 1.3. **Differences between phase separation by the nucleation-growth mechanism and phase separation by the spinodal-decomposition mechanism.**

Nucleation-growth mechanism	Spinodal-decomposition mechanism
The composition of the deposited phase does not vary over time at a constant temperature.	The composition of the two phases varies over time until an equilibrium state is reached.
An interface exists between the two phases; the concentration gradient is fixed.	Initially, the boundary between the two phases is unclear; as phase separation progresses, however, the concentration gradient increases.
Spherical liquid drops are produced that have irregular size and position distribution.	Regularities exist in the size and position distributions of the second phase.
The spherical liquid drops are generated separately and show little mutual contact.	The second phase has a mutually interconnected structure.

per unit volume. Such phase separation differs from crystallization in that whereas the mixing temperature is generally lower than the liquidus temperature, the liquid-liquid interface energy is smaller than the liquid-solid interface energy. Consequently, phase separation rates are generally faster than crystallization rates. The liquid drops grow at a diffusion-governed rate. Until the liquid-drop volume fraction reaches a fixed value, the particle diameter increases in proportion to the one-half power of the time. After the volume fraction has reached a fixed value, however, small particles grow by fusing into large particles so that the particle diameter increases in proportion to the one-third power of the time.[18]

1.3.3.2. Phase Separation by Spinodal Decomposition

The theory of spinodal decomposition was developed by Cahn.[19,20] For a heterogeneous solution where average composition is c_0, but where composition differs slightly from c_0 according to location, a Fourier component of composition is imagined, and the fluctuation in composition is posited as $A \cos \boldsymbol{\beta} \cdot \boldsymbol{\gamma}$ (see Figs. 1.12 and 1.13). The initial rate of spinodal decomposition of such a solution is obtained by solving the following diffusion equation:

$$\frac{\partial c}{\partial t} = M \left(f_0'' \, \nabla^2 c - 2K \, \nabla^4 c \right) \quad . \qquad (1.13)$$

Here, c is the concentration of the solution, ∇c is the concentration gradient, f is the free energy per unit volume, and M is the molecular mobility. The variable K is a gradient energy coefficient and cor-

responds to the surface energy. The following equation is obtained as a solution of Eq. (1.13):

$$c = A \exp \left[R(\beta)t \right] \cos \boldsymbol{\beta} \cdot \boldsymbol{\gamma} \quad . \qquad (1.14)$$

where $R(\beta)$ is an amplification coefficient, $\boldsymbol{\beta}$ is a wave-number vector, and $\boldsymbol{\gamma}$ is a position vector. Here, $R(\beta)$ is given by the following equation:

$$R(\beta) = - \left(f'' + 2K\beta^2 \right) M\beta^2$$

$$= 2KM \left(\beta_c^2 - \beta^2 \right) \beta^2 \quad . \qquad (1.15)$$

In the region where $f'' < 0$, $R(\beta)$ is positive when $\beta < \beta_c = (f''/2K)^{1/2}$. Figure 1.14 shows the relationship between $R(\beta)$ and β, and $R(\beta)$ shows a sharp maximum at $\beta_m = \beta_c / \sqrt{2}$. The numerous wave numbers that existed initially disappear, and as time elapses, the wave number component corresponding to β_m becomes the governing factor.

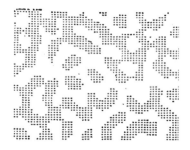

Figure 1.12. Morphology expected from spinodal decomposition (Refs. 19, 21).

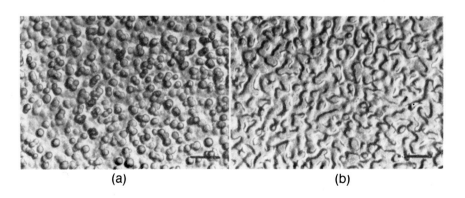

(a)   (b)

Figure 1.13.   Electron micrographs of typical phase-separation structures (Ref. 22) (length of black lines: 0.3 μm).

Cahn's theory was investigated by Zarzycki and Naudin[23] for $PbO-B_2O_3-Al_2O_3$ systems, and by Neilsen[24] and Andreev et al.,[25] for SiO_2-Na_2O glasses, using small-angle x-ray scattering and light scattering. Zarzycki et al., and Neilsen supported Cahn's theory by citing the existence of a linear relationship between the logarithm of the x-ray scattering intensity and the heat treatment time, the existence of a maximum in $R(\beta)$, etc. (see Fig. 1.14). Thus, Cahn's theory concerning spinodal decomposition has been experimentally confirmed. Some phenomena in the very early stages of phase separation do not fit into the theory, and problems regarding some details have persisted, but Cook's correction for thermal fluctuation[26] and Yokota's correction for relaxation[27]

allowed demonstration of the theory in a more complete form.

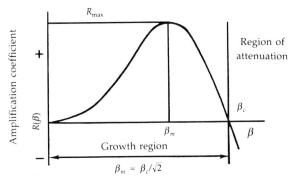

Figure 1.14. Relationship between amplification coefficient and wave number.

1.4. The Glass Transition

1.4.1. Introduction

According to theory, equilibrium structure exists in a liquid and that this structure depends on temperature. In a supercooled liquid, the equilibrium structure changes as the temperature drops; since the relaxation time is short, however, it is thought that the liquid reaches an equilibrium structure at various temperatures. When a supercooled liquid turns into a glass as shown in Fig. 1.1, the viscosity of the glass increases as the temperature drops; as a result, the relaxation time ($\tau = \eta/G$) becomes longer. This eventually results in a lack of time for rearrangement of the internal structure (including the movement of atoms and the rotation of molecules) in response to the change in temperature so that the glass structure is frozen at a certain temperature. After this freezing occurs, only atomic vibrations are possible so that behavior similar to that of a crystal is exhibited. This temperature is T_g and the relaxation time is about 100 seconds. If a glass is maintained at a constant temperature in the vicinity of T_g, the glass structure gradually changes toward an equilibrium structure. It is thought that for each temperature, there is a corresponding equilibrium structure, and conversely, that there is a corresponding temperature for each equilibrium structure. The temperature that corresponds to this structure is called the fictive temperature or structural temperature.

1.4.2. Thermodynamics of the Glass Transition

In addition to P and T, a parameter that expresses the degree of regularity (degree of order) of the system must be introduced to thermodynamically describe the behavior of a glass (or liquid) in the vicinity of T_g.

Figure 1.15 shows the volume-temperature variation (constant pressure) in the glass transition region in idealized form. In Fig. 1.15, AB is the equilibrium volume-temperature curve; D, C, and W have the following respective degrees of order:

$$Z_{T_1}, Z_{T_2}, \text{ and } Z_{T_3} \text{ at } T_1, T_2, \text{ and } T_3 \quad .$$

The corresponding structural tempeatures are $\overline{T}_1$, $\overline{T}_2$, and $\overline{T}_3$.

When a glass is cooled from a certain temperature, the volume departs from curve BA, and follows (for example) curve BCN. Or, if the rate of cooling is faster, the volume will turn at point D and follow BDR. The points Q, M, and W, at the same temperature T_3, have the following degrees of order:

$$Z_{T_1}, Z_{T_2}, \text{ and } Z_{T_3}.$$

The state at M is the same whether the path taken is $D \rightarrow C \rightarrow M$ or $D \rightarrow Q \rightarrow M$. In other words, the thermodynamic state of a glass can be completely expressed by three variables: P, T, and Z.

When C_P' and α' represent the specific heat and expansion coefficient of the vitreous state, respectively; and C_P and α represent the same values in the equilibrium state, the following is posited:

$$\Delta C_P = C_P - C_P' \quad .$$

In Fig. 1.16, the following change is considered:

$$M(P,T,Z_{\overline{T}}) \rightarrow G(P,T + dT, Z_{\overline{T} + d\overline{T}}) \quad .$$

Assuming that the route followed is

$$M(P,T,Z_{\overline{T}}) \rightarrow C(P,\overline{T},Z_{\overline{T}}) \rightarrow D(P,\overline{T} + d\overline{T}, Z_{\overline{T} + d\overline{T}})$$

$$\rightarrow G(P,T + dT, Z_{\overline{T} + d\overline{T}}) \quad ,$$

then

$$dV = V(\alpha' \, dT + \Delta\alpha \, d\overline{T}) \quad ,$$

$$dH(= dQ) = \left(C_P' \, dT + \Delta C_P \, d\overline{T}\right) \quad , \tag{1.16}$$

$$dS = C_P' \frac{dT}{T} + \Delta C_P \frac{d\overline{T}}{\overline{T}} \quad .$$

The increase in entropy is not dQ/T (here, dH/T); the irreversible entropy increase is given by the difference between dS and dH/T.

$$dS = dS_{rev} + dS_{irr}$$

$$dS_{irr} = dS - \frac{dH}{T} \tag{1.17}$$

$$= \Delta C_P \left(\frac{1}{\overline{T}} - \frac{1}{T}\right) d\overline{T} \quad .$$

Because $\Delta C_P > 0$, $d\overline{T} > 0$ or $d\overline{T} < 0$, i.e., $\overline{T}$ increases or decreases, depending on whether $T > \overline{T}$ or $T < \overline{T}$ in Eq. (1.17). In other words, the structural temperature $\overline{T}$ approaches the actual temperature. In Fig. 1.15, if D, which has a structural temperature of $\overline{T}_1$, is maintained at T_3, the structural temperature will approach W, which is the maintenance temperature T_3. Here, only a single parameter was taken, i.e., degree of order Z. However, some doubt remains as to whether a single parameter is sufficient. If we assume that a single parameter is sufficient, then Ehrenfest's Eq. (1.18) should hold true.

$$\Delta\kappa_T \cdot \Delta C_P = TV (\Delta\alpha)^2 \quad . \tag{1.18}$$

However, when Eq. (1.18) is determined for glucose, approximately four values are obtained, not one.[28] This indicates that a single parameter is insufficient. In fact, the need for two or more structural temperatures or relaxation times has been recognized by many researchers, including Ritland[29] and others. However, it is extremely useful to consider the behavior of a glass in the transition region by introducing a single structural temperature $\overline{T}$; however, it is best to think of this as only a first approximation.

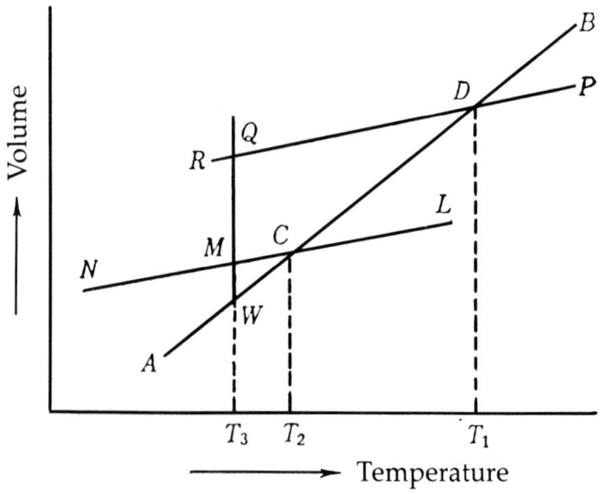

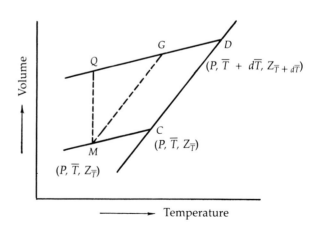

Figure 1.15. Volume–temperature variation of a glass in the transition region.

Figure 1.16. Volume–temperature variation in the transition region expressed by thermodynamic quantities (details in text).

1.4.2.1. Glass-Transition-Rate Theories

In Fig. 1.17, a glass (in a state of equilibrium at B or D) is rapidly taken to the temperature of C or E, and a change of $C \rightarrow F$ or $E \rightarrow F$ occurs. That is to say, a change of $\overline{T}_1 \rightarrow T_2$ or $\overline{T}_3 \rightarrow T_2$ takes place. Here, an equilibrium state is reached more quickly when $\overline{T} > T$ than it is when $\overline{T} < T$. The reason for this is that the volume is greater when $\overline{T} > T$ so that the resistance (viscosity) when equilibrium is approached is smaller.

In an irreversible process, the rate of irreversible entropy increase is expressed as a product of "forces" and "flows" (Ref. 30); in sufficient proximity to a state of equilibrium, both factors are in a proportional relationship.

From Eq. (1.17)

$$\frac{dS_{irr}}{dt} = \Delta C_P \frac{d\overline{T}}{dt} \left(\frac{1}{\overline{T}} - \frac{1}{T} \right) \quad .$$

Here, $\Delta C_P \, (d\overline{T}/dt)$ is a flow, and $[(1/\overline{T}) - (1/T)]$ or $(T - \overline{T})$ is a force. In other words,

$$\frac{d\overline{T}}{dt} = A(T - \overline{T}) \quad , \tag{1.19}$$

where A is a rate constant related to temperature; this can be set equal to the fluidity or reciprocal of the viscosity, and can be expressed as $ae^{-E/kT}$. Thus

$$\frac{d\overline{T}}{dt} = ae^{-E/kT}(T - \overline{T}) \quad . \tag{1.20}$$

As seen in Fig. 1.18, however, the curves approaching $\overline{T}$ are not the same above and below. The reason for this is that E is a function of the volume of the glass, and is thus a function of T and $\overline{T}$. Tool[31] has proposed the following equation:

$$\frac{d\overline{T}}{dt} = ae^{bT + c\overline{T}}(T - \overline{T}) \quad .$$

This equation shows good agreement with experimental values if a, b, and c are appropriately selected.

Tool also proposed the following:

$$\frac{d\tau}{dt} = Ke^{T/g} \cdot e^{\overline{T}/h}(T - \overline{T}) \quad .$$

This equation and the concept of fictive temperature are extremely useful; they allow a convenient description of changes in physical quantities (e.g., changes in density, index of refraction, viscosity, electric conductivity, etc.) in the glass transition region, and also provide useful information about strain relaxation. The index-of-refraction n and strain s relaxation-rate formulas are as follows:

Index of refraction: Collyer's equation

$$\frac{dn}{dt} = Q(n_e - n)e^{n/P} \quad \text{(constant temperature).}$$

Strain relaxation: Tool's equation

$$\frac{-ds}{dt} = K \cdot e^{-s/g} \cdot S \quad \text{(constant temperature).}$$

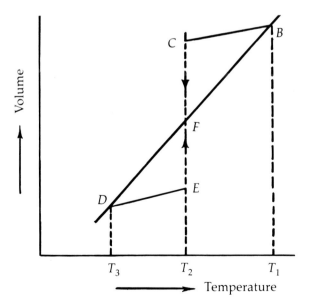

Figure 1.17. Volume variation with abrupt temperature variation.

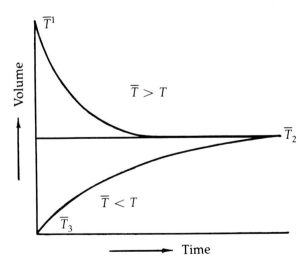

Figure 1.18. Variation of volume over time until the structural temperature T is reached.

References

1. G. S. Smith and G. E. Rindone, *J. Am. Ceram. Soc.* **44**, 72 (1961).
2. W. H. Zacharisen, *J. Am. Chem. Soc.* **54**, 3841 (1932).
3. B. E. Warren, *J. Appl. Phys.* **8**, 645 (1937).
4. E. A. Porai-Koshits in *Proceeding of a Conference on the Structure of Glass* **25**, Leningrad, Russia (1953).
5. H. G. Pfaender and H. Schroeder, *Schott Glaslexikon Moderne* (Verlags GmbH, Munchen 50, Satz, 1980), p. 25.
6. J. C. Slater, *Introduction to Chemical Physics* (McGraw-Hill Book Company, Inc., New York and London, 1939), p. 266.
7. G. Tamman, *J. Soc. Glass Tech.* **9**, 166 (1925).
8. A. Dietzel, *Naturwissenschaften* **23**, 537 (1941); and *Z. Elektrochemie* **48**, 8 (1942).
9. K. Han Sun, *J. Am. Ceram. Soc.* **30**, 277 (1947).
10. L. Pauling, *The Nature of the Chemical Bond* (Kyōritsu Shuppan Co., Ltd., Tokyo, 1962), p. 76.
11. J. C. Phillips, *J. Non-Cryst. Solids* **34**, 167 (1979).
12. T. Izumitani, *Rep. Osaka Ind. Res. Inst.*, No. 311 (June 1958), p. 59.
13. R. Roy, *Symposium on Nucleation and Crystallization in Glass and Melts*, p. 39 (1962).
14. W. Vogel, *Symposium on Nucleation and Crystallization in Glass and Melts*, p. 11 (1962).
15. K. Nakagawa and T. Izumitani, *Phys. Chem. Glasses* **13**, 85 (1972).
16. I. Prigogine, *Bull. Soc. Chim. Belg.* **52**, 115 (1943).
17. J. W. Cahn and R. J. Charles, *Phys. Chem. Glasses* **6**, 186 (1965).
18. Y. Moriya, D. H. Warrington, and R. W. Douglas, *Phys. Chem. Glasses* **8**, 19 (1967).
19. J. W. Cahn, *Acta Metall.* **9**, 795 (1961).
20. J. W. Cahn, *Acta Metall.* **10**, 179 (1962).
21. J. W. Cahn, *J. Chem. Phys.* **42**, 93 (1965).
22. S. Sakka, T. Sakaino, K. Takahashi, Eds., *Glass Handbook* (Asakura Publishing Co., Tokyo, 1975), p. 799.
23. J. Zarzycki and F. Naudin, *J. Non-Cryst. Solids* **1**, 215 (1969).
24. G. N. Neilsen, *Phys. Chem. Glasses* **10**, 54 (1969).
25. N. S. Andreev, G. G. Boiko, and N. A. Bokov, *J. Non-Cryst. Solids* **5**, 41 (1970).
26. H. E. Cook, *Acta Metall.* **18**, 297, (1970).
27. R. Yokota, *Yogyo-Kyokai-Shi (Japan)* **86**, 994, 251–276 (1978); *J. Phys. Soc. Jpn.* **45**, 1, 29–41 (1978).
28. G. O. Jones, *Glass* (Methuen and Co. Ltd., London, 1956), p. 58.
29. H. N. Ritland, *J. Am. Ceram. Soc.* **37**, 370 (1954); and *J. Soc. Glass Tech.* **39**, 99 (1955).
30. G. O. Jones, *Glass* (Methuren and Co., Ltd., London, 1956), p. 63.
31. A. Q. Tool, *J. Am. Ceram. Soc.* **29**, 240 (1946).

CHAPTER 2

Compositions and Physical Properties of Optical Glass

2.1. What is Optical Glass?

2.1.1. Characteristics of Optical Glass

Optical glass refers to glass that is used in lenses and prisms. A special feature of optical glass is that (unlike other types of glass) it is optically homogeneous. Such optical glass was first widely developed by O. Schott, a German, around 1890. From that time on, development has continued with Germany as the center of the optical glass world. Barium crown glass, which was invented by Schott, is famous; it is a well-known fact that the design of color-corrected lenses was made possible by the appearance of this glass. In 1939, however, a new type of optical glass containing lanthanum, thorium, and tantalum was invented by G. W. Morey, an American—with this invention, great advances in the development of high-refraction, low-dispersion glass were achieved. Specifically, the index of refraction jumped from 1.6 to 1.8 at an Abbe number of 50.

The founders of the Japanese optical glass industry were Japan Optical Industries, Ltd. (centered around Mr. Masao Nagaoka) and the Osaka Industrial Experimental Institute (centered around Dr. Toru Takamatsu). After the second world war, research on the compositions of new types of optical glass was conducted mainly by T. Izumitani at the Osaka Industrial Research Institute.

The development of optical glass in Japan focused not only on compositions, but also on manufacturing methods. That is to say, whereas optical glass manufacturing methods introduced from Germany were based almost exclusively on melting in clay pots; borate glasses (containing rare earths) developed after the second world war were melted in platinum pots. As a result, bubbles, and striae in optical glass were greatly reduced. In 1965, continuous melting of optical glass, which might be called the third revolution in the manufacture of optical glass, was achieved by Hoya Glass. In this process, lenses or prisms produced by the feeding of raw materials, melting, pressing, and annealing are continuously fed out of the annealing furnace in the same way that plate glass or bottle glass is processed. Optical glass, which differs from bottle and plate glass in being completely free of striae and bubbles, is produced continuously. This is surely something that was undreamed of until around 1950. In addition to the flourishing Japanese optics industry, the Japanese optical glass industry has also become a world center.

The most important prerequisite for optical glass is homogeneity. It is not an exaggeration to say that all possible technological efforts have been devoted to making this type of glass homogeneous. This is the reason for the stirring, precise annealing, using platinum for melting vessels, etc.—operations that are not used in other glass industries. Some of the properties required in optical glass are as follows:

- The glass must be colorless and transparent.
- The glass must be optically homogeneous and isotropic.
- The glass must be free of bubbles, striae, and strain.
- The glass must have the prescribed optical constants and a small temperature dependence.
- The glass must have sufficient mechanical strength (hardness, elastic modulus, etc.).
- The glass must have a high resistance to weathering.

All of the above properties tie in with the production of homogeneous glass for the purpose of forming images.

Another special feature of optical glass is that many types of optical glasses exist. An examination of the lens of a camera will show that cameras have a large number of lenses, i.e., four, six, or even seven lenses. It is not unusual for zoom lens to contain as many as 13 lenses that are used to eliminate chromatic aberration, spherical aberration, coma aberration, astigmatism, curvature of the image field, and distortion of the image. To eliminate these aberrations, two or more lenses with different indices of refraction must be combined. Chromatic aberration refers to a phenomenon in which images are formed in different positions as a result of different indices of refraction at different wavelengths of light. The other five aberrations are called Seidel's five aberrations, and are defects caused by the spherical shape of the refracting surface. The more types of optical glass that exist, the easier it is to eliminate these aberrations. For example, to eliminate chromatic aberration, the following conditions must be satisfied in lens design:

$$\sum \frac{1}{v_k f_k} = 0 \quad . \tag{2.1}$$

Here, f_k is the focal length of the kth lens and v_k is the Abbe number, a value that corresponds to the reciprocal of the dispersion of the glass. In a case where the lens constitution consists of two lenses (concave and convex),

$$\frac{1}{v_1 f_1} - \frac{1}{v_2 f_2} = 0 \quad ,$$

i.e., $v_1/v_2 = f_2/f_1$. Furthermore, to eliminate curvature of the image field

$$\sum \frac{1}{n_k f_k} = 0 \quad . \tag{2.2}$$

Here, n_k is the index of refraction, in a case where the lens composition consists of two lenses (concave and convex), $1/n_1 f_1 - 1/n_2 f_2 = 0$, i.e., $n_1/n_2 = f_2/f_1$. Accordingly, the following condition must be satisfied to eliminate both curvature of the image field and chromatic aberration at the same time.

$$\frac{n_1}{n_2} = \frac{v_1}{v_2} \quad . \tag{2.3}$$

In general, because the Abbe number v_2 of a concave lens is smaller than the Abbe number v_1 of a convex lens, it is necessary to combine a lens in which $n_1 > n_2$ (that is, in which n_1 and v_1 are both large); that is a high-refraction, low-dispersion convex lens must be combined with a lens in which n_2 and v_2 are both small (a low-refraction, high-dispersion concave lens). For this reason, glasses with various indices of refraction and Abbe numbers are needed; as a result, varieties of optical glass are extremely numerous.

2.1.2. Types and Compositions of Optical Glass

The index of refraction n_d and Abbe number v can be used to classify optical glasses. The term n_d is the index of refraction with respect to the d

line of He (587.6 nm) and is called the mean index of refraction. If the indices of refraction with respect to the F line of hydrogen (486.1 nm) and C line of hydrogen (656.3 nm) are respectively expressed as n_F and n_C, then the Abbe number can be defined as follows: $v = (n_d - 1)/(n_F - n_C)$. Because the index of refraction of a substance differs according to the wavelength of light, dispersion occurs. The degree of light dispersion is expressed as a difference in indices of refraction; $n_F - n_C$ is called the mean dispersion. The Abbe number is proportional to the reciprocal of the mean dispersion so that a small Abbe number indicates a large dispersion.

Although optical glasses are classified according to n_d and v, such glasses can be roughly divided into crown glasses and flint glasses. In crown glasses, the value of v is 55 or greater; in flint glasses, the value of v is 50 or less. Furthermore, crown glasses can be further subdivided into BK, K, SK, etc., whereas flint glasses can be classified as F, LF, SF, etc. These designations are formed by adding the names of specific chemical components to the general names crown and flint; the adjective "light" or "heavy" is also added according to the index of refraction.

The designator BK refers to borosilicate crown glass; BK 7 is produced in larger quantities than any other optical glass. The designator SK refers to glass that contains BaO and has a high specific gravity. This type of glass was invented by Schott; compared with the crown and flint glasses of the time, it offered high refraction and low dispersion, and was therefore useful in lens design. Furthermore, flint glasses contain PbO; LF refers to light flint glasses containing relatively small amounts of PbO, while SF conversely refers to heavy flint glasses containing large amounts of PbO (some of the latter glasses containing 70% or more PbO). Figure 2.1 shows types and names of optical glasses classified according to n_d and v.

The designator PKS refers to glasses that use P_2O_5 as a glass-forming oxide to reduce dispersion; FK refers to glasses in which the index of refraction is lowered by adding fluorides; and KzF refers to glasses in which partial dispersion in the long-wavelength region is increased by adding B_2O_3. Kurz flint (KzF) refers to the fact that partial dispersion in the short-wavelength region is smaller in such glasses than it is in ordinary flint glasses containing PbO. LaK and LaF are new types of glass that came into existence after the second world war. These glasses contain La_2O_3, and were developed to satisfy design requirements for high refraction and low dispersion.

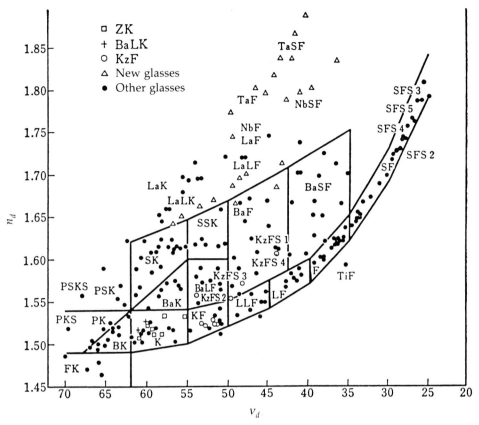

Figure 2.1. Optical glasses from the Hoya Glass Catalog.

However, a type of glass that satisfies other requirements for low refraction and high dispersion is TiF; a special feature of glasses of this type is that they contain TiO_2 and fluorides. In conventional glasses, as seen in Fig. 2.1, v decreases with an increase in n_d so that dispersion tends to increase. Conversely, new types of glass have been manufactured in such a way so as to achieve the opposite relationship.

Thus, more than 200 types of optical glass exist; however, these glasses can be classified relatively simply in terms of composition. As shown in Table 2.1 and Fig. 2.2, the glasses consist of approximately twelve series.

The following sections are an outline description of the individual series listed in Table 2.1 and shown in Fig. 2.2.

① SF-F-LF-KF and K Series

This series corresponds to ① in Fig. 2.2. Glasses in this series consist of SiO_2–PbO–R_2O. As PbO is replaced by SiO_2, there is a shift from SF to KF; the dispersion decreases with a decrease in PbO, and the index of refraction also decreases. In the case of K, PbO is no longer present, but is replaced by BaO or CaO.

Heavy flint glasses contain large amounts of PbO; as a result, such glasses are easy to color, have high specific gravity, low melting temperature, and poor "acid resistance." Because the alkali increases as PbO decreases, there is a deterioration in water resistance so that surface deterioration tends to occur in KF, etc. Furthermore, with the increase in SiO_2, LF and KF show a higher melting temperature and become difficult to melt. This series includes a special heavy flint glass, SFS, that contains TiO_2 and has a high partial dispersion. Furthermore, SiO_2–R_2O–BaO–TiO_2–Nb_2O_5 systems have been developed to improve the hardness of heavy flint glasses.

② BaSF–BaF–BaLF and BaK–BaLK Series

This series consists of glasses of series ① in which PbO is replaced by BaO; such glasses consist of SiO_2–PbO–BaO–R_2O systems. Moving into the crown region, PbO disappears, and BaO and the alkali are increased. As PbO is replaced by BaO, the Abbe number increases, the ease of coloration decreases, the softening point Sp rises, and the expansion coefficient decreases. Because of their BaO content, BaSF–BaF glasses have poor acid resistance compared with SF-F glasses. In

Table 2.1.   Principal components of optical glasses.

Series	Flint	Crown
①	SiO_2–PbO–R_2O	→SiO_2–RO–R_2O
②	SiO_2–PbO–BaO–R_2O	→SiO_2–BaO–R_2O
②′	SiO_2–B_2O_3–PbO–BaO	→SiO_2–B_2O_3–BaO
③	(SiO_2)–B_2O_3–La_2O_3–PbO–Al_2O_3	→(SiO_2)–B_2O_3–La_2O_3–RO–ZrO_2
③′	(SiO_2)–B_2O_3–La_2O_3–PbO–RO	→(SiO_2)–B_2O_3–La_2O_3–RO–ZrO_2
④	(SiO_2)–B_2O_3–La_2O_3–ZnO–TiO_2–ZrO_2	→(SiO_2)–B_2O_3–La_2O_3–ZnO–Nb_2O_5
⑤	B_2O_3–La_2O_3–Gd_2O_3–Y_2O_3–Ta_2O_5	→B_2O_3–La_2O_3–Gd_2O_3–Y_2O_3
⑥		SiO_2–B_2O_3–R_2O–BaO
⑦		P_2O_5–$(Al_2O_3$–$B_2O_3)$–R_2O–BaO
⑧		SiO_2–B_2O_3–K_2O–KF
⑨	SiO_2–TiO_2–KF	
⑩	SiO_2–B_2O_3–R_2O–Sb_2O_3	
⑪	B_2O_3–(Al_2O_3)–PbO–RO	
⑫		SiO_2–R_2O–ZnO

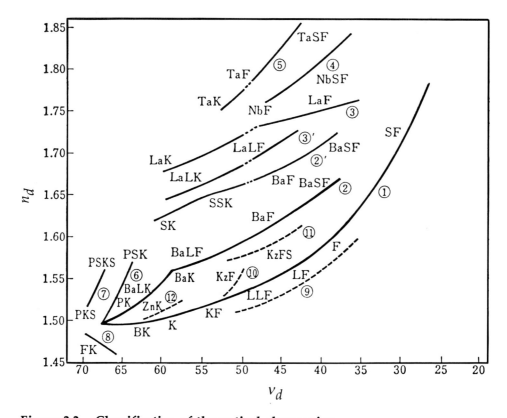

Figure 2.2.   Classification of the optical glass series.

BaK–BaLK glasses, no PbO is present, and some BaO is replaced by alkali. Water resistance deteriorates as the alkali content increases.

②′ BaSF–BaF–SSK–SK Series

There are two types of BaSF and BaF glasses. Specifically, BaSF and BaF glasses in series ② contain an alkali, whereas BaSF and BaF glasses in this series ②′ contain no alkali, but rather contain B_2O_3 as an easy melting component. In other words, series ②′ consists of SiO_2–B_2O_3–PbO–BaO systems. In the case of SSK–SK, PbO disappears completely; they are SiO_2–B_2O_3–BaO systems and are appropriate for the manufacture of high-refraction, low-dispersion glasses. The replacement of SiO_2 by B_2O_3 is an effective means of increasing the Abbe number. Because SK glasses include large amounts of B_2O_3 and BaO, their chemical durability is poor, and staining tends to occur. Glasses formed by adding TiO_2 to this system make up the SK 16–SK 18–SSK 5–BaF 10–BaSF 7–BaSF 8 group, the members of this group have the highest indices of refraction of any silicate glasses. In some of the glasses in this group, chemical durability has been improved by replacing B_2O_3 and PbO with Li_2O and TiO_2. Furthermore, in the case of SK 16, chemical durability has been considerably improved by the introduction of La_2O_3 and ZrO_2.

③ LaF–LaK Series

This series consists of new optical glasses that appeared after the second world war; their basic composition is B_2O_3–La_2O_3–RO. The LaK glasses contain ZnO, CaO, SrO, and BaO as RO; LaF glasses contain PbO or TiO_2 in addition to these RO components. These glasses have a low viscosity at the time of melting so that devitrification tends to occur easily. Accordingly, ZrO_2, SiO_2, etc., are added as antidevitrification agents; CdO and ThO_2, which were used in the past, caused environmental problems and are therefore no longer used. Because the SiO_2 component is small, the acid resistance of these glasses is extremely poor. In regard to water resistance, LaK glasses (containing alkali earths) are inferior. Because La_2O_3 dissolves in water only with difficulty, it has a beneficial effect on water resistance.

③′ LaLF–LaLK Series

This series consists of B_2O_3–La_2O_3–RO glasses, which are intermediate between series ③ and series ②′. These glasses are formed by replacing La_2O_3 in series ③ glasses with an RO component so that the amount of La_2O_3 is decreased, and the RO component is increased. Because these glasses are basically borate glasses with a large RO component, they, along with KzFS glasses, are among the poorest in terms of chemical durability. The LaLK group includes glasses that have especially poor water resistance.

④ NbSF–NbF Series

These glasses consist of B_2O_3–La_2O_3–RO–Nb_2O_5 systems. Previously CdO, which has a great antidevitrification effect, was included as RO, but now ZnO has come into use. Besides SiO_2 and ZrO_2, NbSF glasses contain TiO_2 and WO_3 as antidevitrification agents. The acid resistance of these glasses is also relatively good, and the water resistance is extremely good.

⑤ TaSF–TaF–TaK Series

This series consists of high-refraction, low-dispersion glasses that presently occupy the highest positions on the n_d–v graph; ThO_2 was previously used in these glasses, but has been replaced by (Gd_2O_3, Y_2O_3) so that the glasses now consist of B_2O_3–La_2O_3–(Gd_2O_3, Y_2O_3)–(Ta_2O_5, Nb_2O_5) systems. A special feature of TaSF glasses is that the glass-forming oxide B_2O_3 and RO component are greatly reduced to achieve a high index of refraction. The liquidus temperature is lowered by the eutectic melting of La_2O_3, Gd_2O_3, Y_2O_3, and Ta_2O_5 so that the glasses are stabilized against devitrification. However, because the glasses contain only a small RO component, they are still somewhat unstable with respect to devitrification. The glasses also contain only a small amount of SiO_2. However, because they contain large amounts of La_2O_3, ZrO_2, and Ta_2O_5, they are among the most superior of all optical glasses in terms of water resistance and compare favorably with the silicate glasses in series ②. Another special feature of glasses in this group is that they have an extremely high hardness. Both TaF and TaK glasses are created by reducing the amount of Ta_2O_5 in TaSF glasses, replacing it with Gd_2O_3 and Y_2O_3, and increasing the amount of the glass-forming oxide B_2O_3. This TaSF–TaF–TaK series forms the highest n_d–v line; however, this n_d–v line is determined by the types of constituent ions and the vitrification region. Replacing B_2O_3 with SiO_2 increases the solubility of high-valence ions; in this way, glasses with an n_d of 1.85 or greater were developed.

⑥ PSK–PK and BK Series

To obtain a high Abbe number and a low index of refraction, it is necessary to increase the amount of the glass-forming oxide as much as possible, to decrease the amount of the modifier oxide, and to primarily use a monovalent ion that gives a low index of refraction for the modifier oxide. If the glass-forming oxide SiO_2 is used in large amounts, the glass becomes difficult to melt. Accordingly, an easily meltable glass-forming oxide B_2O_3 is introduced; in this way, a minimal-refraction minimal-dispersion borosilicate glass SiO_2–B_2O_3–R_2O is obtained. Such glasses are called phosphate crown PK; it is important to note, however, that the composition contains no P_2O_5 and that the glasses are borosilicate glasses. These PK glasses contain slightly more B_2O_3 than BK glasses; PSK glasses contain more BaO than PK glasses. The glass BK 7 is the most commonly used type of optical glass and is the most widely produced stable optical glass.

⑦ PSKS–PKS Series

Special phosphate-crown PKS refers to phosphate glasses consisting of P_2O_5–R_2O–RO. Both B_2O_3 and Al_2O_3 are added in order to improve chemical durability. Special heavy-phosphate-crown (PSKS) glasses are obtained by replacing R_2O with RO. Phosphate glasses are used for the following reason: the cationic field of P^{5+} is strong; as a result, the polarizability of O^{2-} is small. Accordingly, low-refraction low-dispersion glasses are obtained. Because P_2O_5 is soluble in water, glasses of this type have an extremely poor water resistance.

⑧ FK Series

Glasses in this series consist of SiO_2–B_2O_3–K_2O–KF systems; Al_2O_3 is added as an antidevitrification agent. The fact that the F^- ion has a smaller ionic radius and smaller polarizability than the O^{2-} ion is used to achieve a low refraction and low dispersion. Because fluorides are ionic, these glasses have a poor acid resistance.

⑨ TiSF–TiF–TiK Series

This series consists of SiO_2–TiO_2–KF systems; the glasses in this series are low-refraction, high-dispersion glasses. TiO_2 increases the dispersion more than PbO. At the same time, the introduction of F^- in place of O^{2-} results in a low refraction. Because these glasses contain a fluoride, parallel striae tend to occur.

⑩ KzF Series

In this series, Sb_2O_3 is introduced in place of the PbO in series ① so that the glasses consist of SiO_2–B_2O_3–R_2O–Sb_2O_3 systems. Because the absorption of Sb^{3+} in the ultraviolet region is at a shorter wavelength than that of Pb^{2+}, partial dispersion in the blue portion of the visible spectrum is small. Furthermore, because the glasses contain B_2O_3, they are susceptible to the effects of near-infrared absorption so that partial dispersion in the red portion of the visible spectrum is large.[1] Thus, these glasses provide so-called "abnormal partial dispersion." Because the glasses contain B_2O_3 and alkali, their water resistance is extremely poor.

⑪ KzFS Series

This series consists of B_2O_3–PbO glasses containing no Sb_2O_3. Like KzF glasses, these glasses show abnormal partial dispersion. These glasses have the poorest water and acid resistance of any optical glasses; however, this is understandable when we consider that such glasses are based on a B_2O_3–PbO composition.

⑫ ZnK Series

These are SiO_2–R_2O–ZnO glasses; in these glasses, ZnO is used as an RO component in place of the PbO of K (crown) glasses or BaO of BaK glasses. Because Zn^{2+} is a divalent ion with a small ionic radius, its polarizing effect is great, and a low-refraction, low-dispersion glass is obtained. Because these glasses contain ZnO, they have a small expansion coefficient and good water resistance.

Above, we discussed the relationship between types and compositions of optical glasses. Recently, a fluorophosphate optical glass FCD 10 has been developed that has an n_d of 1.45650 and a v of 90.8. This glass consists of a P_2O_5–AlF_3–RF_2 system, and shows an abnormally large partial dispersion in the short-wavelength region. Glasses that show a large abnormal partial dispersion in the short-wavelength region include the above mentioned low-refraction low-dispersion fluorophosphate glass, fluorosilicate glasses, and phosphate crown glasses. Glasses that show a large abnormal partial dispersion in the long-wavelength region include lanthanum glasses containing B_2O_3 and kurz flint glasses. The compositions and physical and chemical characteristics of representative glasses are shown in Tables 2.2 and 2.3.

Table 2.2. Compositions of optical glasses (wt%).

	SiO_2	B_2O_3	Al_2O_3	Na_2O	K_2O	CaO	ZnO	BaO	PbO	Sb_2O_3	As_2O_3	KHF_2	TiO_2	P_2O_5
SF 2	40.9			0.5	6.8				50.8		1.0			
SF 6	26.9			0.5	1.0				71.3		0.3			
F 2	45.7			3.6	5.0				45.1		0.6			
F 8	50.2			3.8	5.6				39.7		0.3			
LF 1	54.3			4.4	7.8		1.0	1.5	34.9		0.8			
LF 7	33.9			2.5	7.9				45.1		0.6			
KF 2	66.7			15.9			3.5		12.9		1.0			
K 8	70.7	2.8		5.8	11.3			6.4	2.8		0.4			
BaSF 2	23.6	10.5	1.5			7.0	5.9	29.3	14.2	0.4	0.6		7.0	
BaF 10	30.9	9.2	0.3			4.0	5.3	41.3	4.6		0.3		3.6	
BaLF 1	53.8			1.5	9.5		10.0	14.2	10.7		0.3			
BaK 1	47.7	4.2	1.0	1.0	7.5		8.6	29.0			0.2			
BaK 2	59.6	3.0		3.0	10.0		4.8	19.0			0.6			
SSK 2	37.3	6.3	2.5				8.0	39.7	5.2		1.0			
SK 1	40.1	5.7	2.5				8.5	42.2	0.5		0.5			
SK 3	35.0	11.9	4.5					45.9	0.6	1.6	1.0			
SK 5	38.7	14.9	5.0					40.1		0.3	1.0			
SK 16	30.8	17.9	1.4	0.3				48.7		0.4	0.5			
PK 1	68.2	13.5	1.3		12.1			0.5	3.0				2.4	
BK 1	71.4	6.5		5.2	13.9	2.0					1.0			
BK 7	68.9	10.1		8.8	8.4		MgO	2.8			1.0			
PKS 1		4.0	9.0		11.6		4.0				1.0			70.4
FK 1	51.0	18.3	8.3		7.3						0.2	14.4		
FK 3	47.7	17.4	1.4	2.2	2.4						0.3	16.0		
KzF 1	46.0	14.0	3.0		12.0		4.0			20.6	0.4			
ZK 1	55.7	7.0		1.0	16.0		20.0				0.3			

	SiO_2	B_2O_3	La_2O_3	ZrO_2	CaO	BaO	PbO
LaF 2	4	32.7	29.0	7.5	11.0		15.8
LaF 3	4	37.3	25.7	7.4	10.7	4.0	10.7
LaK 10		41.3	32.4	8.1	12.1		6.1

Table 2.3. Physical and chemical characteristics of optical glasses.[a,b]

	n_d	v_d	Thermal properties		Chemical durability[c]	
			Sp (°C)	$\alpha \times 10^7$ (°C^{-1})	Da	Dw
SF 2	1.64769	33.9	483	98	3	2
SF 6	1.80518	25.5	480	85	4	2
F 2	1.62004	36.3	454	101	1	2
F 8	1.59551	39.2	481	85	1	1
LF 1	1.57309	42.7	471	99	1	2
LF 7	1.57501	41.3	471	94	1	2
KF 2	1.52630	51.0	514	94	1	2
K 8	1.51276	59.8	542	88	1	3
BaSF 2	1.66446	35.9	536	85	3	2
BaF 10	1.67003	47.2	671	82	4	2
BaLF 1						
BaK 1	1.57250	57.5	639	80	4	1
BaK 2	1.53996	59.7	618	83	1	2
SSK 2	1.62230	53.1	680	74	5a[d]	2
SK 1	1.61025	56.5	680	76	4	2
SK 3	1.60881	58.9	689	76	5a[d]	2
SK 5	1.58913	61.2	680	72	4	2
SK 16	1.62041	60.3	672	75	5b[d]	3
PK 1						
BK 1	1.51009	63.4	610	92	1	2
BK 7	1.51680	64.2	642	88	1	1
PKS 1	1.51728	69.6	545	80	2	5
FK 1	1.47069	67.2	492	87	4	2
FK 3	1.46450	65.8	475	87	4	2
KzF 1	1.55115	49.6	525	71	4	2
ZK 1	1.53315	58.1	592	78	4	2
LaF 2	1.74400	44.9	584	77	5a	2
LaF 3	1.71700	47.9	668	93	5a	1
LaK 10	1.72000	50.3	650	74	5a	2

Grade	Da 1	Da 2	Da 3	Da 4	Da 5
Weight reduction (%)	<0.03	0.03 to 0.10	0.10 to 0.30	0.30 to 1.00	>1.00

Grade	Dw 1	Dw 2	Dw 3	Dw 4
Weight reduction (%)	<0.05	0.05 to 0.20	0.20 to 0.50	0.50 to 1.00

[a] Acid resistance was measured by immersing specimens for 1 hour at 50°C in 150 ml of 1/100 N HNO_3; the specimens were classified according to percentage of weight reduction.

[b] Durability was measured by boiling the coarsely pulverized glass for a fixed period of time in a fixed amount of distilled water, and then measuring the weight reduction (%).

[c] New methods of measurement and data analysis are noted in Section 2.2.3 and Table 2.10.

[d] 5a is better than 5b.

2.2.   How is the Composition of an Optical Glass Determined?

In Chapter 1, we talked about what glass is, and about what kinds of substances can form glasses. In this section, we need to talk about how the composition of an optical glass is put together. The point of departure for such a discussion is a study of physical properties: namely, how the index of refraction and dispersion is determined. Once the relationship between refractive index and glass components is understood, it is necessary to learn about the vitrification region to determine the composition. Next, moving ahead from the simple composition determined by the vitrification region, it is necessary to find a way of stabilizing the glass by modifying the composition so that devitrification will not occur in mass production. In addition, it is necessary to consider the composition from the standpoint of improving chemical durability so that dimming and staining will not occur when the glass is worked into lenses or prisms. The composition of an optical glass is determined through such a process.

2.2.1.   Refraction and Dispersion

The first question we must ask is, what determines the index of refraction and the dispersion of a glass? The relationship between a substance and its refractive index is given by the Lorentz-Lorenz formula:

$$\frac{n^2 - 1}{n^2 + 2} \cdot V = \frac{4}{3} \pi N \alpha = R \quad . \qquad (2.4)$$

Here, n is the index of refraction, V is the molecular volume, N is Avogadro's number, α is the polarizability, and R is called the molecular refraction. The variable R is a quantity that is peculiar to the substance involved and is independent of temperature, pressure, or state of aggregation. When n is close to 1, R is given by the Gladstone-Dale formula:

$$n - 1 = \frac{R}{V} \quad . \qquad (2.5)$$

The index of refraction increases with an increase in molecular refraction and with a decrease in molecular volume.

2.2.1.1.   Molecular Refraction

The molecular refraction of a glass is expressed as the sum of ionic refractions, but is determined principally by oxygen ions, which have the largest ionic refraction. Kordes[2] has pointed out that although the ionic refraction of a cation is fixed, the ionic refraction of an oxygen ion depends on its bonding state. The oxygen ions in glass include bridging oxygens that are bound to two Si as indicated by –Si–O–Si–, and non-bridging oxygens (–Si–O$^-$M$^+$) that are bound to an Si and to a modifier cation. As shown in Table 2.4, the ionic refraction of a singly-bound oxygen is greater than the ionic refraction of a bridging oxygen.

The ionic refraction of oxygen ions varies according to the type of network-forming oxide (network former) used, the amount of modifier oxide used, and the polarizing effect of the modifier cation, etc.[3]

Relationship between $R_{O^{2-}}$ and the Network-Forming Ion. The $R_{O^{2-}}$ decreases with an increase in the charge of the network-forming cation, and increases with an increase in the ionic radius of the modifier cation (see Table 2.5).

Table 2.4.   Ionic refractions of a few ions.

Ionic refraction	Stevels	Biltz	Kordes	Free gas state
$R_{O^{2-}}$ (bound oxygen)	3.66	3.70	3.67	6.95
$R_{O^{2-}}$ (singly-bound oxygen)	4.78	4.91		
R_{K^+}		2.95		2.25
R_{Na^+}		0.84		0.47
$R_{Ca^{2+}}$	1.55			1.40

Table 2.5.   Relationship between ionic refraction of oxygen ion and glass-forming oxide.

$(XO_4)^{n-}$	n/2 Mg^{2+}	n/2 Ca^{2+}	n/2 Sr^{2+}	n/2 Ba^{2+}	nK$^+$
$(BO_3)^{3-}$	3.79				
$(SiO_4)^{4-}$	3.83	4.53	4.67	4.97	
$(PO_4)^{3-}$		3.89			
$(SO_4)^{6-}$		3.5	3.41	3.50	3.60
$(ClO_4)^-$				3.26	

Relationship between $R_{O^{2-}}$ and the Amount of Modifier Oxide. As the O/Si ratio increases, i.e., as the number of nonbridging oxygens increases as a result of the introduction of a modifier oxide, the ionic refraction $R_{O^{2-}}$ of a single-bond oxygen ion increases (see Table 2.6).

Relationship between $R_{O^{2-}}$ and the Modifier Ion. In an orthosilicate, $R_{O^{2-}}$ increases with a decrease in the field strength, z/a^2, of the modifier cation (Table 2.7). The reason for this is as follows: as the size of the cation increases, the polarizing effect decreases so that the polarizability of the oxygen ion increases. Accordingly, $R_{O^{2-}}$ increases.

Relationship between $R_{O^{2-}}$ and the Outer-Shell Electron Structure of the Cation. Cations may be divided into noble-gas type ions that have 8 electrons in the outer shell, and non-noble-gas type ions that have 18 electrons. However, the electronic structure of a non-noble-gas type cation is influenced by the oxygen anion. Especially in the case of the Pb^{2+} ion, which has an $(18 + 2)$ electronic structure, the O^{2-} ion causes loosening of the electron shell so that an ionic refraction greater than that of the O^{2-} ion is produced.

2.2.1.2. Molecular Volume

The molecular volume of a glass is also determined by the way in which the oxygen ions that have the largest ionic radius are packed. In systems containing low-valence cations, the volume that contains one mole of O^{2-} increases with an increase in the ionic radius of the cation, or with a decrease in the cationic field strength z/a^2 (Ref. 4). The order is as follows: $Cs^+ > K^+ > Na^+ > Li^+$, $Ba^{2+} > Sr^{2+} > Ca^{2+}$ (for example). This order occurs for the following reasons: the size of the cation contributes to the volume, and the adjacent O^{2-} ions are attracted by the cationic field. In the case of high-valence ions, however, the number of surrounding anions, i.e., the coordination number,

increases with an increase in the ionic radius of the cation so that the molecular volume decreases. For example, this variation occurs in the following order: $Th^{4+} < Zr^{4+} < Ti^{4+}$, $Ta^{5+} < Nb^{5+} < V^{5+}$. In other words, there are some cases in which the molecular volume increases and other cases in which the molecular volume decreases, with an increase in the size of the modifier cation contained in the system. That is to say, the molecular volume is determined by the cationic field strength in the case of low-valence ions and by the coordination number of the cation in the case of high-valence ions. The degree of packing may be thought of as the number of O^{2-} ions per unit volume.

2.2.1.3. Refractive Index

As mentioned above, the index of refraction is given by the following equation: $n - 1 = R/V$. The index of refraction may be increased either by increasing the polarizability of the O^{2-} ions, or by increasing the number of oxygen ions per unit volume. Figure 2.3 shows the relationship of the ionic radius to the molecular refraction and molecular volume in various groups of the periodic table. As Fig. 2.3 shows, the molecular refraction and molecular volume both increase with an increase in ionic radius in the case of divalent and trivalent ions. Furthermore, the index of refraction increases. Accordingly, the index of refraction is determined by the polarizability of the oxygen ions in the case of glasses containing low-valence ions. For glasses containing high-valence ions, i.e., tetravalent, pentavalent, or hexavalent ions, both the molecular refraction and the molecular volume decrease with an increase in ionic radius, and the index of refraction increases. Accordingly, it is seen that the index of refraction is determined by the degree of packing of O^{2-} ions in the case of such glasses. In other words, the polarizing effect of the cation governs the index of refraction in glasses containing low-valence ions, whereas the coordination number of the cation determines the index of refraction in glasses containing high-valence ions.

Table 2.6. Relationship between ionic refraction of oxygen ion and O:Si ratio.

	O/Si	$R_{O^{2-}}$ (crystal)	$R_{O^{2-}}$ (glass)
SiO_2	2	3.52	3.66
$BaSiO_5$	2.5	4.09	
$BaSi_2O_8$	2.66	4.28	
$BaSiO_3$	3.0	4.46	5.5
$BaSiO_4$	4.0	4.97	

Table 2.7. Relationship between ionic refraction of oxygen ion and cationic field strength, z/a^2.

	$2Be^{2+}$	Si^{4+}	$4/3 Al^{3+}$	$2Mg^{2+}$	$2Ca^{2+}$	$2Sr^{2+}$	$2Ba^{2+}$
$R_{O^{2-}}$	3.35	3.62	3.75	3.83	4.5	4.67	4.97
z/a^2	0.86	1.57	0.84	0.45	0.33	0.28	0.24

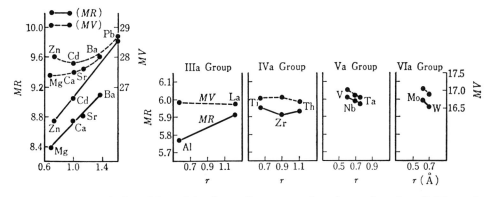

Figure 2.3. Relationship of ionic radius to molecular refraction (*MR*) and molecular volume (*MV*) within the same group.

2.2.1.4. Dispersion

Absorption and Dispersion. Dispersion is a phenomenon that occurs because the index of refraction differs according to the wavelength of light. The relationship between index of refraction and frequency of light is given (for example) by the Drude-Voigt dispersion equation:

$$n - 1 = \frac{N_1 e^2}{2\pi m} \sum \frac{f_i}{v_{0i}^2 - v^2} \quad . \tag{2.6}$$

Here, N_1 is the number of oxygen ions per unit volume, e and m are the charge and mass of an electron, v_0 is the characteristic frequency of absorption of the oxygen ion, f_i is the oscillator strength, and v is the frequency of the light. As shown in Fig. 2.4, there are ordinarily two or three ultraviolet absorptions and one infrared absorption. In the vicinity of wavelengths where an absorption occurs the index of refraction increases abruptly as indicated by the above equation, and then gradually decreases as the wavelength becomes longer. From Eq. (2.6), the index of refraction n_d, mean dispersion $n_F - n_C$, and Abbe number can be expressed as follows:

$$n_d - 1 = \frac{N_1 e^2}{2\pi m} \sum \frac{f}{v_0^2 - v_d^2} \quad , \tag{2.7}$$

$$n_F - n_C = \frac{N_1 e^2}{2\pi m} (v_F^2 - v_C^2)$$

$$\sum \frac{f}{(v_0^2 - v_F^2)(v_0^2 - v_C^2)} \quad . \tag{2.8}$$

Abbe number =

$$\frac{n_d - 1}{n_F - n_C} = \frac{(v_0^2 - v_F^2)(v_0^2 - v_C^2)}{(v_F^2 - v_C^2)(v_0^2 - v_d^2)} \quad . \tag{2.9}$$

Both the index of refraction and the mean dispersion are functions of N_1, f, and v_0. However, the mean dispersion $n_F - n_C$ has two terms involving v_0^2 in the denominator and is more strongly influenced by ultraviolet absorption wavelengths. In other words, whereas the index of refraction is a function of degree of packing, oscillator strength, and absorption wavelength; dispersion may be thought of as being determined mainly by absorption wavelength. Furthermore, dispersion will be expected to increase as ultraviolet absorption occurs at longer wavelengths. Furthermore, it is seen that as an approximation the Abbe number is a function of ultraviolet absorption wavelength alone, and that the Abbe number increases with increasing v_0. This has been confirmed by measuring ultraviolet absorption using glass films less than 1 μm thick.[5]

As shown in Fig. 2.5, the mean dispersion increases with increasing absorption wavelength. Because the absorptions of glasses containing Th^{4+} and Ta^{5+} are at short wavelengths, these glasses have a low dispersion. Moreover, the glasses have a high index of refraction. This indicates that the degree of packing is the cause of the high refraction.

Effect of the Cation on Ultraviolet Absorption. Let us consider the effect of modifier cations in glasses on ultraviolet absorption.[6] Assuming that

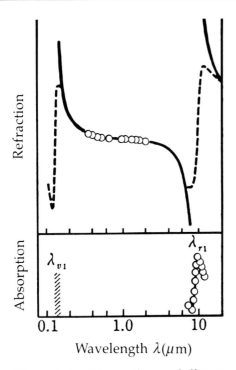

Figure 2.4. Absorption and dispersion in glass.

the dispersion in the Drude-Voigt Eq. (2.6) is determined only by one ultraviolet absorption (principally the ultraviolet absorption of nonbridging oxygens), the following equation is obtained:

$$n^2 - 1 = \left(\frac{e^2}{\pi m C^2}\right) \frac{N f_0}{\left(1/\lambda_0^2 - 1/\lambda^2\right)} \quad . \qquad (2.10)$$

Taking the reciprocal of this

$$\frac{1}{n^2 - 1} = \frac{\pi m c^2}{e^2 N f_0}\left(\frac{1}{\lambda_0^2} - \frac{1}{\lambda^2}\right) \quad . \qquad (2.11)$$

Because N can be calculated from $N_0 \cdot \rho/M$, λ_0 and f_0 can be determined from the linear relationship in the above equation. Figure 2.6 shows the relationship between ultraviolet absorption wavelength λ_0 and ions in various groups and rows of the periodic table; Fig. 2.7 shows the relationship between oscillator strength f_0 and said ions.

It is very interesting to note that among the ions that impart a high index of refraction, Pb^{2+}, Ti^{4+}, and Nb^{5+} also produce a high dispersion and have absorption at long wavelengths; whereas Ba^{2+}, La^{3+}, Y^{3+}, and Th^{4+} give a low dispersion and show a high f_0. Furthermore, whereas the absorption wavelengths λ_0 of low-valence ions (alkalies, alkali earths) move toward the long-

wavelength side as the ionic radius of the cation increases, the absorption wavelengths of high-valence ions (Ti^{4+}, Th^{4+}, Ta^{5+}, etc.) move toward the short-wavelength side as the ionic radius increases. This may be interpreted as follows: as the ionic radius increases, the coordination number increases so that the polarizability of the oxygen decreases. Accordingly, the characteristic absorption moves toward the short-wavelength side. There is also a tendency for λ_0 to move to the long-wavelength side as the valence increases, but the reason for this is not understood.

A tendency is observed for the oscillator strength f_0 to decrease with an increase in the cationic field strength z/a^2. This is thought to result from a decrease in the polarizability of nonbridging oxygens so that the shell electrons are tightened. Figure 2.8 shows the relationship between f_0 and λ_0. In the case of low-valence ions, the oscillator strength increases as the absorption wavelength moves toward the long-wavelength side; in the case of high-valence ions, however, the oscillator strength tends to decrease as the absorption wavelength moves toward the long-wavelength side. If the oscillator strength f_0 is thought of as the number of available electrons contributing to dispersion, then we understand that the number of available electrons contributing to dispersion would increase as the absorption of nonbridging oxygens moved toward longer wavelengths. In the case of high-valence ions, however, we do not understand why the number of electrons contributing to dispersion would increase in spite of the fact that ultraviolet absorption moves toward shorter wavelengths with an increase in the radius of the cation.

Cases Where there are Three Ultraviolet Absorptions and One Infrared Absorption. In optical glasses, the following absorptions occur in the ultraviolet region[7]:
- Absorption by bridging oxygen ions.
- Absorption by nonbridging oxygen ions.
- Absorption by cations (in cases where the glass contains lead ions, etc.).

Furthermore, there is also an absorption in the infrared region. Accordingly, the dispersion equation can be written as follows:

$$n^2 - 1 = \frac{KN_1 f_1}{1/\lambda_1^2 - 1/\lambda^2} + \frac{KN_2 f_2}{1/\lambda_2^2 - 1/\lambda^2}$$

$$+ \frac{KN_3 f_3}{1/\lambda_3^2 - 1/\lambda^2} - \frac{KN_4 f_4}{1/\lambda^2 - 1/\lambda_4^2} \quad . \qquad (2.12)$$

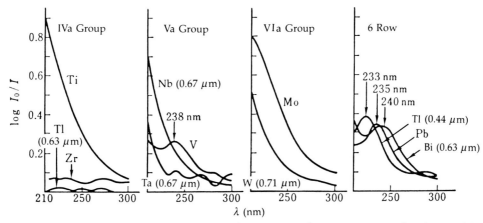

Figure 2.5.   Absorption of $BO_{1.5}$ (80%), BaO (18%), and MO_n (2%) glasses in the ultraviolet region.

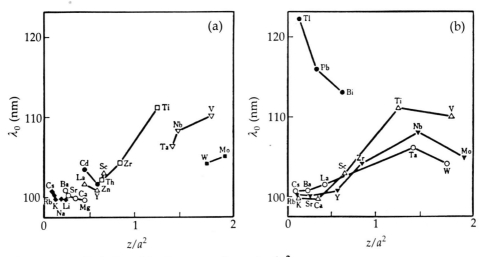

Figure 2.6.   Relationship between λ_0 and z/a^2.

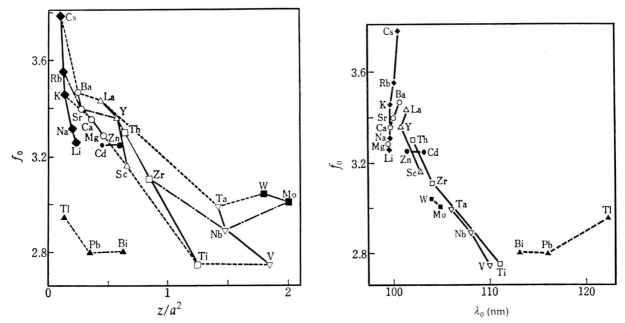

Figure 2.7.   Relationship between f_0 and z/a^2.   **Figure 2.8.   Relationship between f_0 and λ_0.**

The first term arises from absorption by bridging oxygen ions; λ_1 is the peak wavelength and $KN_1 f_1$ is the intensity. Similarly, the second term arises from absorption by nonbridging oxygen ions, and the third term arises from absorption by cations (e.g., lead). (In some cases, therefore, there is no third term.) Finally, the fourth term arises from infrared absorption.

The wavelengths and intensities of these characteristic absorptions can be determined by substituting the index of refraction from the t line to the i line into this equation. For example, the values of $KN_i f_i$ and λ_i for normal-dispersion glasses K 7 and F 2 are shown in Table 2.8. Here, the λ_1 values of 0.106 μm and 0.122 μm, and the λ_2 value of 0.172 μm for K 7, and the λ_1 values of 0.106 μm and 0.122 μm in the λ_2 value of 0.181 μm and the λ_3 values of 0.207 μm and 0.251 μm for F 2 are actually measured values.

Comparing K 7 and F 2 glasses in terms of the data shown in Table 2.8, we see that the replacement of SiO_2 by PbO results in a slight decrease in the first absorption intensity, i.e., absorption intensity by bridging oxygen ions decreases, whereas absorption intensity by nonbridging oxygen ions increases. At the same time, absorption by Pb ions appears at 0.207 and 0.251 μm. Not much change is seen in infrared absorption. It is interesting that the respective absorption intensities show the following relationship: $KN_1 f_1 > KN_2 f_2 > KN_3 f_3 > KN_4 f_4$. Figure 2.9 shows how $n_d - 1$, $n_F - n_C$, and v_d vary when $KN_1 f_1$, $KN_2 f_2$, $KN_3 f_3$, and $KN_4 f_4$ are varied.

As $KN_1 f_1$ increases, $n_d - 1$ also increases, but $n_F - n_C$ shows only a very slight change; accordingly, v_d increases. As $KN_2 f_2$ increases, both $n_d - 1$ and $n_F - n_C$ increase; as a result, v_d decreases. As $KN_3 f_3$ increases, $n_F - n_C$ increases remarkably; accordingly, v_d shows a remarkable decrease. An increase in $KN_4 f_4$ causes somewhat an increase in $n_F - n_C$, and a corresponding decrease in v_d. Figure 2.10 shows how $n_d - 1$,

$n_F - n_C$, and v_d changes when λ_1, λ_2, λ_3, and λ_4 vary. As λ_1 increases, $n_d - 1$ and $n_F - n_C$ do not show a linear relationship; as a result, v_d first increases, and then again decreases. As λ_2 and λ_3 increase, $n_d - 1$ and $n_F - n_C$ increase, and $n_F - n_C$ shows an especially conspicuous increase; as a result, v_d decreases. The wavelength λ_4 does not have any great effect.

Vacuum Ultraviolet Absorption in Optical Glasses. As described above, the dispersion of a glass is determined by absorption in the ultraviolet and infrared regions. However, absorption in the ultraviolet region has only been estimated from the dispersion equation; almost no actual measurements have been performed. The reason that such absorption has not been measured despite the fact that it constitutes extremely important data is that the absorption wavelengths are in the vacuum ultraviolet region so that the absorption coefficients involved are large, i.e., 10^5 to 10^6 cm^{-1}; as a result, it has been difficult to measure the spectroscopic transmittance curve. Instead of absorption spectra, Hirota and Izumitani[8] have measured reflection spectra in the 13 to 6 eV (90 to 210 nm) region using a Johnson-Onaka vacuum spectrometer. A multiline hydrogen spectrum produced by hydrogen discharge tube was used as a light source; the light was dispersed by diffraction-grating vacuum monochromator and was directed onto the glass specimen at an angle of incidence of 10 deg. The reflected light was received by a sodium salicylate fluorescent plate, and the fluorescence of this plate was detected by a photomultiplier. The degree of vacuum was 5×10^{-3} Torr (Fig. 2.11). Transmission in the vacuum ultraviolet region was measured using 2.0- to 3.5-μm films formed by immersing a quartz tube in the glass melt and blowing. The experimental results are described in the following sections.

Reflection Spectra of Silicate Glasses. Figure 2.12 shows reflection spectra of SiO_2–Na_2O

Table 2.8. Values of $KN_i f_i$ and λ_i for normal-dispersion glasses K 7 and F 2.

	n_d	v_d	λ_1	$KN_1 f_1$	λ_2	$KN_2 f_2$	λ_3	$KN_3 f_3$	λ_4	$KN_4 f_4$
K 7	1.51112	60.41	0.0684	151.40	0.172	1.925	—	—	9.8	0.00825
			0.106	19.10						
			0.122	18.27						
F 2	1.62004	36.37	0.0684	158.62	0.181	9.944	0.207	1.043	9.8	0.00840
			0.106	17.38			0.251	0.399		
			0.122	14.89						

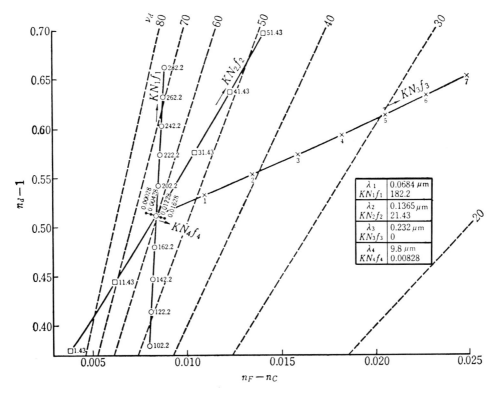

Figure 2.9. Effect of *KNf* on *n* and *v*.

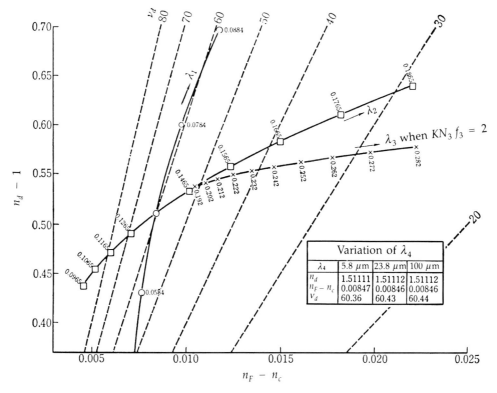

Figure 2.10. Effect of λ on *n* and *v*.

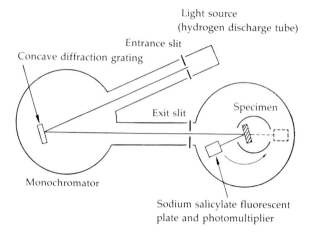

Figure 2.11. Johnson-Onaka vacuum spectrometer.

binary glasses. In fused silica, sharp absorption bands exist at 11.6 eV (107 nm) and 10.4 eV (120 nm). This spectrum shows fairly good agreement with measurement results of Philipp[9] and Sigel.[10] The absorption band at 11.6 eV is thought to be an interband transition from the bonding level to the antibonding level characteristic of SiO_4 tetrahedrons.[10] Furthermore, the absorption band at 10.2 eV is thought to be a Wannier exciton transition.[11] As Na_2O is added, the intensities of these two absorption bands gradually drop, and the bands are shifted slightly toward longer wavelengths. When Na_2O reaches 33.3 mol%, an absorption band of nonbridging oxygens bound to Na^+ appears faintly in the vicinity of 7.5 eV; when Na_2O reaches 36.5 mol%, this band becomes clear. The addition of Na_2O causes the appearance of absorption by nonbridging oxygens; however, a special feature of this absorption is that it is extremely weak compared with absorption by bridging oxygens.

Figure 2.13 shows reflection spectra that are due to the replacement of Na_2O by Li_2O with the amount of alkali oxide at 36.5 mol%. The two systems show extremely similar spectra.

Reflection spectra obtained when SiO_2, Na_2O, and K_2O were held constant and the RO component was varied are shown in Fig. 2.14. In the case

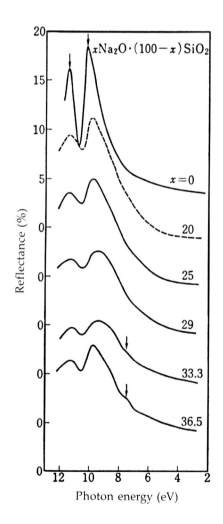

Figure 2.12.   Reflection spectra of SiO_2–Na_2O binary glasses (vacuum ultraviolet to visible region).

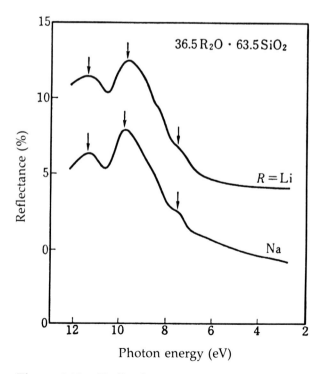

Figure 2.13.   Reflection spectra obtained when alkali is replaced.

of CaO, SrO, and BaO, a single band is generated in addition to the two short-wavelength absorptions related to SiO_2 and the alkali-related absorption. The peak of this band shifts toward longer wavelengths in the order CaO → SrO → BaO. Of the two absorptions on the long-wavelength side, it is thought that the absorption in the vicinity of 9 to 8.4 eV (138 to 148 nm) is caused by oxygen ions singly bound to the alkali, and that the absorption in the vicinity of 7.5 to 6.9 eV (166 to 180 nm) is caused by oxygen ions singly bound to alkali-earth ions. In the case of PbO, the situation is different: absorptions occur at 5.85 eV (212 nm) and 4.81 eV (258 nm). It is thought that these absorptions are caused by intra-ion transitions of Pb^{2+}, i.e., that the absorption band at 5.85 eV is caused by a $^1S_0 \rightarrow {}^1P_1$ transition, and the absorption band at 4.81 eV by a $^1S_0 \rightarrow {}^3P_1$ transition.[12,13]

Figure 2.15 shows reflection spectra for silicate glasses containing Al_2O_3. A Wannier exciton band absorption moves from 9.8 eV toward lower energies as the amount of Al_2O_3 is increased. This indicates that Al_2O_3 acts as a glass-forming oxide.

Figures 2.16, 2.17, and 2.18, respectively, show reflection spectra for systems containing trivalent Y^{3+} and La^{3+}; for systems containing tetravalent Ti^{4+}, Zr^{4+}, and Th^{4+}; and for systems containing pentavalent Nb^{5+} and Ta^{5+}. Absorptions attributable to Ti^{4+}, Zr^{4+}, and Th^{4+} ions were seen at 5.3, 7.49, and 8.88 eV; this indicates that absorptions by nonbridging oxygen to these ions move toward higher energies as the ionic radius increases. Furthermore, absorption by oxygens singly bound to Nb^{5+} and Ta^{5+} appeared at 5.87 and 6.6 eV, showing a shift toward higher energies with an increase in ionic radius.

Transmission and Absorption Spectra of Borate Glasses Containing Various Cations. Figures 2.19 and 2.20 show reflection spectra and transmission spectra of B_2O_3–BaO binary systems. In Fig. 2.19, one absorption band is observed at 8.9 eV (140 nm), and another band-like entity is

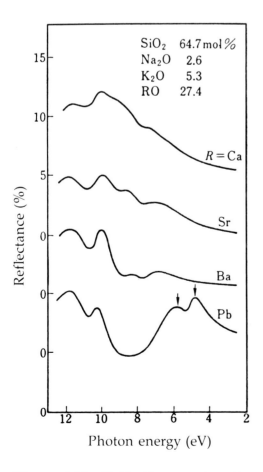

Figure 2.14. Reflection spectra obtained when alkali earths are replaced.

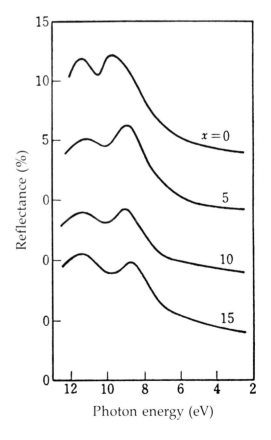

Figure 2.15. Reflection spectra for a silicate glass containing X% Al_2O_3.

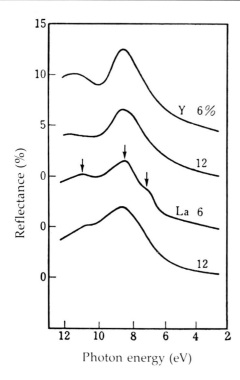

Figure 2.16. Reflection spectra for systems containing Y^{3+} and La^{3+}.

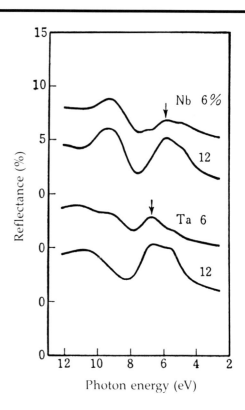

Figure 2.18. Reflection spectra for systems containing pentavalent Nb^{5+} and Ta^{5+}.

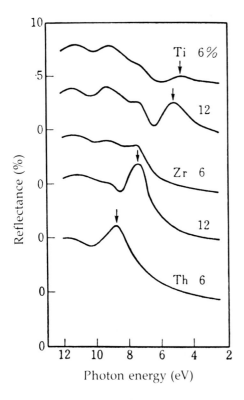

Figure 2.17. Reflection spectra for systems containing tetravalent Ti^{4+}, Zr^{4+}, and Th^{4+}.

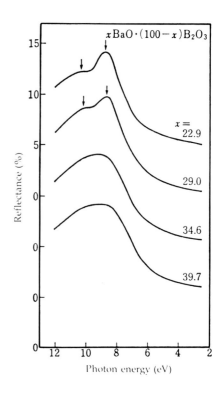

Figure 2.19. Reflectance spectra of B_2O_3–BaO binary glasses.

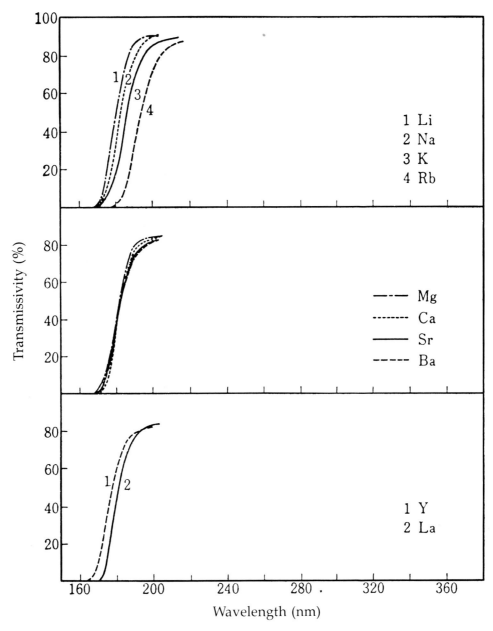

Figure 2.20.   Transmission spectra of 10BaO–5R$_{m/n}$O–85B$_{2/3}$O glasses (1).

present in the vicinity of 10.4 eV (119 nm); these bands broaden with an increase in the amount of BaO. Figure 2.20 shows transmissivities obtained when various cations from the periodic table were substituted into a basic glass consisting of 10BaO, 5R$_{m/n}$O, 85B$_{2/3}$O. In groups Ia, IIa and IIIa, the transmission spectra are fairly steep. In group Ia, there is a shift toward longer wavelengths in the order Li → Na → K → Rb, i.e., with increasing ionic radius. In group IIa, there is very little change; however, a tendency toward a shift to longer wavelengths exists with an increase in

ionic radius. In group IIIa as well, La is shifted toward longer wavelengths than Y. These results show a good correspondence with λ_0 determined from a single-term dispersion equation (see Fig. 2.6).

In groups IVa, Va, and VIa, λ_0 determined from the aforementioned dispersion equation moves toward longer wavelengths as the ionic radius decreases so that the tendency seen here is the opposite of that observed in the case of low-valence ions. Figure 2.21 shows transmission spectra of glasses containing group IVa and group Va ions. The absorption bands are shifted conspicuously toward longer wavelengths as the ionic radius decreases: Th → Zr → Ti, Ta → Nb → V. These results show good agreement with results obtained from the dispersion equation.

In group IIb, Zn and Cd show behavior similar to that of group IIa ions; compared with group IIa ions, however, they show a greater shift toward longer wavelengths (Fig. 2.22).

Next, by decreasing the amount of added ions, we determined absorption peaks for glasses showing high absorption and broad transmission curves. We investigated group IVa, Va, and VIa ions in detail using a $10.8BaO$, $0.2R_{m/n}O$, $89B_{2/3}O$ glass as a basic glass. Absorption coefficients were calculated from $T = (1 - R)^2 e^{-kd}$ and thicknesses were determined from interference fringes. As a result, we found in the case of Ti that a broad absorption occurs that has a peak in the vicinity of 6.3 eV (198 nm) (see Fig. 2.23). We saw a similar tendency in the case of Zr, but this tendency was not as clear. The absorption coefficient of a glass with 0.4% added $Ti_{1/2}O$ was approximately

4400 cm^{-1}. In group Va, we also found that the absorption of a glass containing Nb had a peak at 6.8 eV (182 nm) (see Fig. 2.24). This absorption did not change even in the case of melting in a reducing atmosphere. The absorption of a glass containing V^{5+} was found to be in the vicinity of 6.2 eV (201 nm). Results for group VIa ions are shown in Fig. 2.25. We confirmed that the glass containing Mo^{6+} showed absorption at a longer wavelength than the glass containing W^{6+}.

Measurement results that we obtained by decreasing the amount of added ions in glasses containing row 6b ions (Tl^+, Pb^{2+}, Bi^{3+}) are shown in Fig. 2.26. In these glasses, we found that a clear absorption band is present in the vicinity of 5.0 to 5.8 eV, and that an absorption band is also present in the vicinity of 6.7 eV. The absorption on the long-wavelength side shifted toward longer wavelengths in the order Tl → Pb → Bi; the absorption peak was a 5.64 eV (220 nm) in the case of Tl^+, 5.40 eV (230 nm) in the case of Pb^{2+}, and 5.22 eV (238 nm) in the case of Bi^{3+}. In addition, the absorption in the vicinity of 6.7 eV also

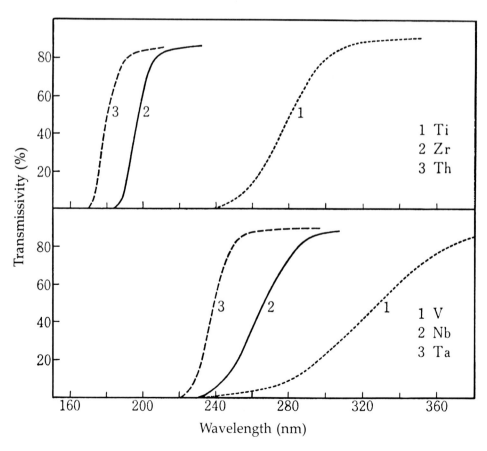

Figure 2.21.   Transmission spectra of $10BaO-5R_{m/n}O-85B_{2/3}O$ glasses (2).

showed a peak that shifted toward longer wavelengths in the order of Tl^+, Pb^{2+}, Bi^{3+}; the respective peaks were at 6.82 eV (182 nm), 6.70 eV (185 nm), and 6.61 eV (188 nm). These absorptions are thought to result from intra-ion transitions in the ions themselves; the bands in the vicinity of

6.7 eV are thought to result from a $^1S_0 \rightarrow ^1P_1$ transition, whereas the longer-wavelength bands are thought to result from a $^1S_0 \rightarrow ^3P_1$ transition. These two absorptions move toward longer wavelengths as the optical basicity of the glass increases.

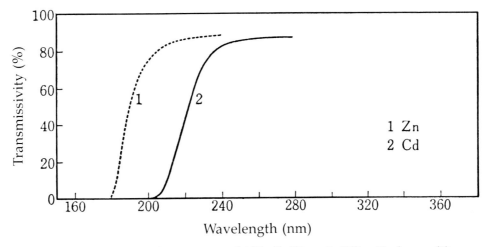

Figure 2.22.   Transmission spectra of $10BaO-5R_{m/n}O-85B_{2/3}O$ glasses (3).

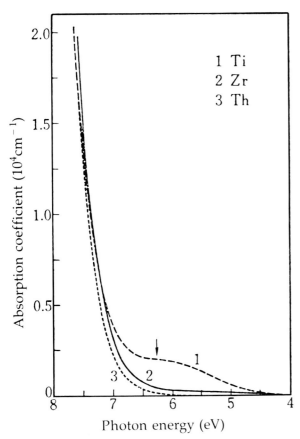

Figure 2.23. Absorption spectra of 10.8BaO–$0.2R_{m/n}O-89B_{2/3}O$ glasses (1).

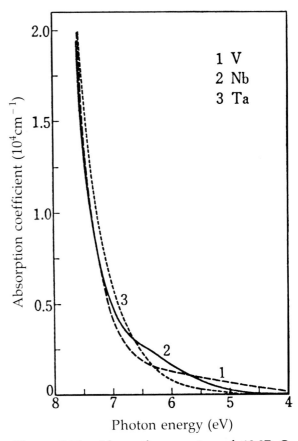

Figure 2.24. Absorption spectra of 10.8BaO–$0.2R_{m/n}O-89B_{2/3}O$ glasses (2).

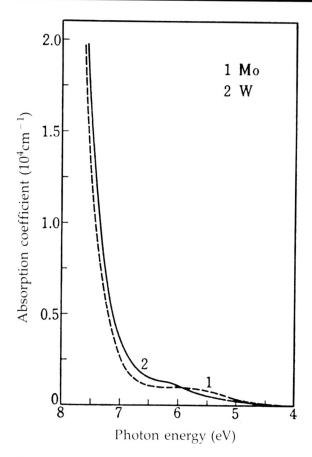

Figure 2.25. Absorption spectra of 10.8BaO–0.2R$_{m/n}$O–89B$_{2/3}$O glasses (3).

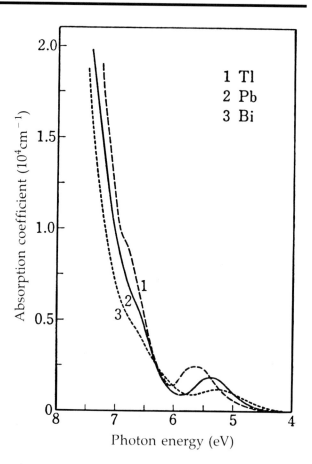

Figure 2.26. Absorption spectra of 10.8BaO–0.2R$_{m/n}$O–89B$_{2/3}$O glasses (4).

Absorption in Phosphate Glasses and Fluorophosphate Glasses. Figure 2.27 shows reflection spectra of P_2O_5–BaO glasses. In these glasses, a single band was found at 9.1 to 9.7 eV (136 to 128 nm). For aluminophosphate glasses and borophosphate glasses, absorptions by oxygen ions bound to Al^{3+} and B^{3+} were found at 8.6 eV and 9.1 eV, respectively, in addition to an absorption by P_2O_5 at 9.7 eV (see Fig. 2.28). This figure shows that Al is present as a modifier ion in phosphate glasses (unlike the case of silicate glasses). Figure 2.29 shows reflection spectra obtained when the amount of P_2O_5 was decreased in fluorophosphate glasses. These results indicate that the absorption band at 11.2 eV (111 nm) is caused by $[AlF_6]^{3-}$.

Through reflection spectra and transmission spectra of silicate, borate, and phosphate glasses, the ultraviolet absorption of glasses (which has been an unknown world until now) has begun to reveal its full form. The locations of absorptions have become clear for nonbridging oxygen ions bound to various modifier ions (and for some types of cations themselves); these locations show

agreement with orderings predicted from the dispersion equation. Specifically, the fact that an increase in ionic radius results in a shift of absorption to longer wavelengths in the case of low-valence ions and a shift to shorter wavelengths in the case of high-valence ions has been experimentally confirmed (though with insufficient theoretical support); this may be viewed as a great achievement. In the future, it will also become possible to express the optical dispersion equation on actually measured absorption wavelengths and absorption coefficients (oscillator strengths), allowing us to obtain a physical basis so that great advances may be expected in the theory of dispersion in optical glasses.

2.2.2. Vitrification Region and Devitrification

2.2.2.1. Vitrification Region

Once high-refraction, high-dispersion components (Ti^{4+}, Pb^{2+}, etc.), high-refraction, low-dispersion components (La^{3+}, Gd^{3+}, Y^{3+}, Zr^{4+},

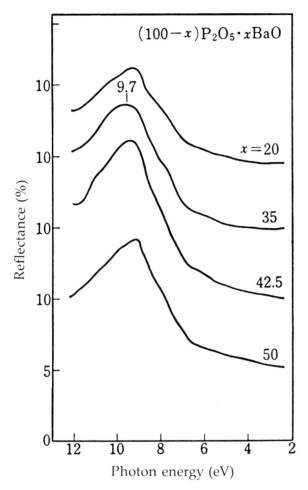

Figure 2.27. Reflection spectra of P_2O_5–BaO binary glasses.

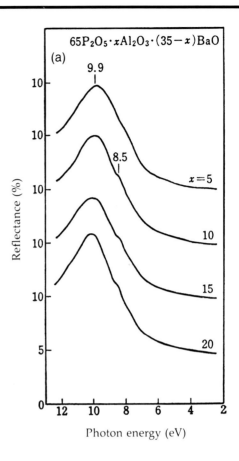

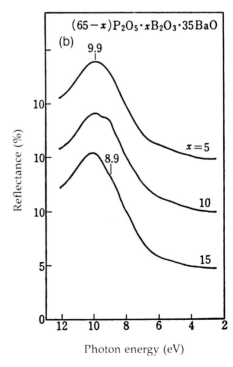

Figure 2.28. (a) Reflection spectra of aluminophosphate glasses and (b) reflection spectra of borophosphate glasses.

Ta^{5+}, etc.), and low-refraction, low-dispersion components (P_2O_5, F^-, etc.) are known from research described in the previous section, we must determine the vitrification region by making triangular diagrams of these modifier ions and glass-forming components, and alkali or alkali-earth components. Then, the composition that offers the greatest stability and the desired optical constants (e.g., high-refraction, low-dispersion or high-refraction, high-dispersion) is selected from within the vitrification region. It is no exaggeration to say that the development of new optical glasses depends on the selection of these three components, and on the discovery of new vitrification regions.

Fortunately, in Japan, we have the broadly based research of Imaoka[14] concerning vitrification regions. However, because the amounts of glass used in the experiments are small, it is best to consider somewhat restricted vitrification regions in cases where such data is used for industrial purposes. Next, let us describe the vitrification regions that include the basic compositions of optical glasses. The ternary system that forms the basis of flint glasses is SiO_2–R_2O–PbO; whereas the ternary system that forms the basis of SK glasses is SiO_2–B_2O_3–RO. Furthermore, the basic system of La glasses is B_2O_3–RO–La_2O_3. These vitrification regions are limited by the phase separation limit (point A), vitrification limit (point B), and solubility limit (point C) (see Fig. 2.30). The vitrification limit B is the point above which a glass network cannot maintain continuity; point C is the point where solubility reaches saturation. Both B and C may be thought of as points where vitrification becomes impossible because of devitrification. To increase the index of refraction, it is necessary to decrease the amount of glass-forming oxide as far as possible. If this is decreased too far, however, the solubility limit is reached, and devitrification occurs.

Figure 2.31 shows vitrification regions obtained when various oxides were introduced into a B_2O_3–BaO system. A tendency is seen for the vitrification region to expand with a decrease in cationic field strength.

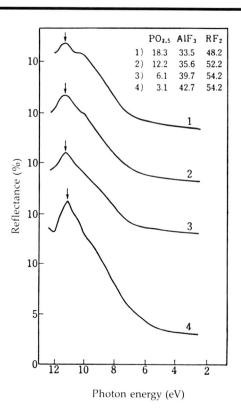

Figure 2.29.   Reflection spectra of fluorophosphate glasses.

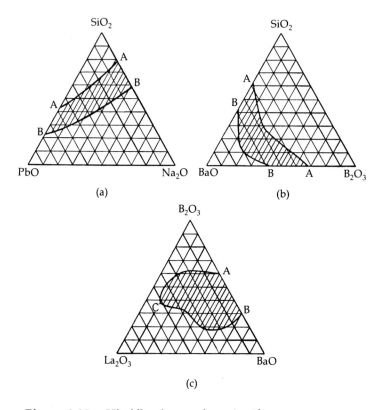

Figure 2.30.   Vitrification regions (wt%).

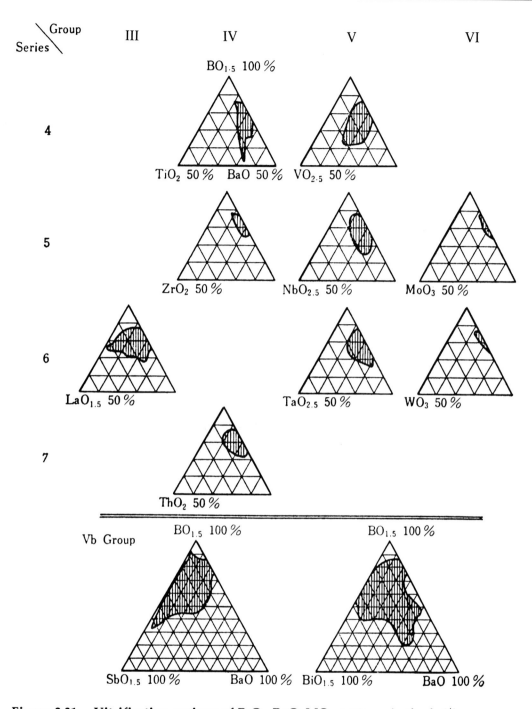

Figure 2.31.   Vitrification regions of B_2O_3–BaO–MO_n systems (cationic %).

As explained in Chapter 1 of Ref. 15, the mixing of two components is determined as follows:

$$w = \left(E_{AB} - \frac{E_{AA} + E_{BB}}{2} \right) > 0$$

(tendency toward separation).

$$w = \left(E_{AB} - \frac{E_{AA} + E_{BB}}{2} \right) < 0$$

(tendency toward homogenization).

If bond energy is approximated by the magnitude of the cationic field, then immiscibility occurs when both components have a high cationic field strength z/a^2. As the difference in cationic field strength increases, the two components become miscible.

In a ternary system,[16] miscibility is determined as follows:

$$\frac{\partial T_C}{\partial x}$$

$$= -\frac{1}{2R}\frac{(\alpha_{AB} - \alpha_{BC} + \alpha_{AC})(\alpha_{AB} + \alpha_{BC} - \alpha_{AC})}{\alpha_{AB}} .$$

$$(2.13)$$

When $\alpha_{AB} < |\alpha_{AC} - \alpha_{BC}|$, the tendency toward phase separation of AB is increased by the C component; when $\alpha_{AB} > |\alpha_{AC} - \alpha_{BC}|$, the tendency toward phase separation is decreased by the C component.

In the case of B_2O_3–BaO–MO_n, the tendency of B_2O_3 and MO_n to separate is alleviated by BaO. Systems in which there is a great difference in field strength between the cation and B_2O_3 show less tendency toward phase separation and appear to be stabilized.

To increase the index of refraction, it is desirable to exclude alkali or divalent components. However, high-valence ions (e.g., La^{3+}, Th^{4+}, Ta^{5+}, etc.) by themselves will not mix with glass-forming oxides (e.g., B_2O_3). Furthermore, two high-valence ions will not mix with each other. If a third high-valence ion is introduced, however, homogenization occurs, and the liquidus temperature drops so that vitrification occurs. This is seen in the B_2O_3–La_2O_3–Ta_2O_5–ThO_2 and ZrO_2 systems shown in Fig. 2.32. This also appears to be a case

in which the condition for homogenization $\alpha_{AB} > |\alpha_{AC} - \alpha_{BC}|$ is satisfied by three components with similar cationic fields (La, Ta, Th) being placed together so that mixing occurs. This is an interesting phenomenon and is also useful information for producing a high-refractive glass; such a phenomenon has also been seen in SiO_2–TiO_2–NaF systems.

2.2.2.2. Devitrification

Even when glass components have mixed to form a homogeneous solution, devitrification may occur. Devitrification refers to the growth of crystals from the glass so that transparency is lost. It is the same thing as crystallization. The crystallization of glass was discussed in Chapter 1; the protection of glass against devitrification is the first condition that must be satisfied when glass is manufactured industrially. To stabilize a glass against devitrification, its composition is finely regulated. Here, the discovery of antidevitrification agents acquires an important significance for the stabilization of such glasses. Because the liquidus temperature and viscosity have a connection with crystal formation and growth, antidevitrification agents consist of components that increase the viscosity at the liquidus temperature, or components that lower the liquidus temperature. Components that lower the liquidus temperature depend on the basic system of the glass involved; a specific component may not lower the liquidus temperature in every system. Furthermore, optimal amounts for lowering the liquidus temperature exist. Among glass compositions that provide given optical constants, we may say that the glass which is most stable with respect to devitrification is the glass that has the

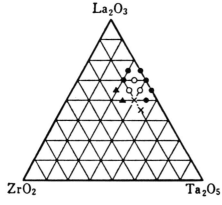
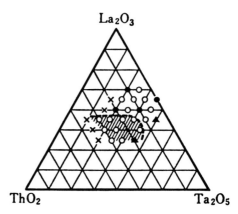

Figure 2.32. Vitrification regions of $20B_2O_3$–La_2O_3–Ta_2O_5–ThO_2 and ZrO_2 systems (wt%).

best composition. For example, changes in the liquidus temperature and softening point that occurred when various components were added to a 31% B_2O_3, 42% CdO, 27% La_2O_3 glass are shown in Fig. 2.33. From this experiment, SiO_2 was found to be a component that lowers the liquidus temperature and also increases the viscosity; and TiO_2, ZrO_2 (<2%), and Al_2O_3 were found to be components that lower the liquidus temperature, but do not cause any great change in the viscosity. Because the addition of Al_2O_3 lowers the index of refraction, such an addition is undesirable in this

case. In the end, the composition improvements shown in Table 2.9 were achieved.

Among these antidevitrification components, TiO_2 has been used in the past as a nucleation agent for crystallization. The sudden decrease in devitrification caused by the addition of 2% TiO_2 is one of the unforgettable memories of the author's youth.

In the case of lanthanum borate glasses, SiO_2 and ZrO_2 are generally used as viscosity-increasing components, and TiO_2 and WO_3 as components that lower the liquidus temperature.

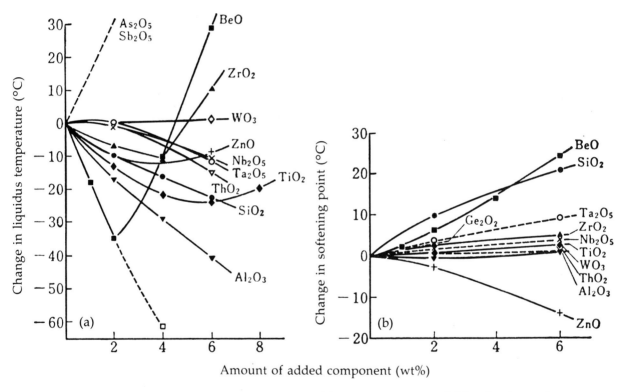

Figure 2.33. Effects of added components: (a) liquidus temperature and (b) softening point.

Table 2.9. Composition improvements (wt %).

	B_2O_3	CdO	La_2O_3	SiO_2	ZrO_2	TiO_2	T_l (°C)	Sp (°C)	log η_{Tl}	$T_{l\,min}$[a]
1	31	42	27				981	636	0.52	637
2	28.5	42	27	2.5			975	636	0.55	651
3	26	42	27	5			969	642	0.63	671
4	26	40	27	5	2		962	649	0.75	682
5	26	38	27	5	4		955	656	0.87	691
6	26	37	27	5	4	1	945			
7	26	36	27	5	4	2	942	660	1.02	701

[a] $T_{l\,min}$ is the minimum temperature for crystal growth.

The liquidus temperature was measured as follows: 10 to 20 mesh glass particles were placed in bucket-type open cells (2-mm-long diameter, 1-mm-short diameter, 0.7-mm depth) formed at 5-mm intervals in a platinum plate ($\sim$2-cm width, 22-cm length). This plate was placed in a furnace with a well-controlled temperature gradient and was heated for 30 minutes so that a sufficient equilibrium was reached. Afterward, the specimen was allowed to cool by standing in air. This experiment was repeated, and the liquidus temperature was determined from the maximum temperature at which crystals were able to exist.

The development of this devitrification test method has made it possible to ascertain the stability of a glass using approximately 20 g of molten glass. As a result, it has become possible to ascertain the most stable glass without the test melting of 200 kg of glass. Thus, a method for developing new optical glasses has been established by the author:

- Ascertaining the relationship of glass components to the index of refraction and Abbe number.
- Forming a three-component vitrification region.
- Measuring the liquidus temperature of the selected basic glass composition using a devitrification plate.
- Determining the softening point.
- Deciding on the most stable glass.

Optical glasses are beginning to be manufactured in Japan as well, and we have managed to advance beyond the mere imitation of German Schott glass.

2.2.3. Chemical Durability of Glasses

Chemical durability[17] is something that must be given careful consideration even after a stable glass composition has been determined. The reason for this is that optical glasses are polished to convert such glasses into prisms and lenses. In glass cold-working processes, glasses come into contact principally with water and sometimes with alkali; the glasses are also exposed to erosion by these substances.

Staining occurs when a glass comes into contact with large amounts of water; dimming occurs when a glass comes into contact with water vapor or mist in the atmosphere. The durability of a glass with respect to these phemonema must be evaluated. First, however, we need to discuss the kinds of chemical reactions that take place between glass and water, and the essential nature

and formation mechanisms of staining, dimming, and latent scratching.

2.2.3.1.   Reaction between Glass and Water

Basically, two reactions occur between a glass and water. One is dissolution (erosion) of the glass network; the other is an ion exchange reaction between the modifier cations of the glass and H_3O^+ ions.

Water dissociates into hydrogen and hydroxy ions.

$$H_2O \rightleftharpoons H^+ \left(H_3O^+\right) + OH^- .$$

The respective dissociated ions react with the glass in the following material.

Breaking of the Glass Network.

$$-O-\overset{|}{\underset{|}{Si}}-O-\overset{|}{\underset{|}{Si}}-O + Na^+OH^-$$

$$\rightarrow -O-\overset{|}{\underset{|}{Si}}-OH + NaO-\overset{|}{\underset{|}{Si}}-O- .$$

The destroyed glass network forms Na^+, etc., and small molecules such as Na_2SiO_3, etc., in the solution, and these undergo dissociative dissolution.

Selective Leaching of Modifier Ions. Modifier ions (alkali, etc.) in the glass undergo an exchange reaction with dissociated H^+ ions in the water; the H^+ ions enter holes (interstitial spaces) in the glass network so that a hydrated layer or silica-gel-like layer is formed, and cations in the glass are leached into the water (Fig. 2.34).

2.2.3.2.   Formation Mechanism of Staining

As described above, an ion exchange reaction takes place between modifier ions and H_3O^+ ions (Fig. 2.35). As a result, a layer with a low index of refraction, in which modifier ions have decreased and H_3O^+ ions have infiltrated, is formed on the surface of the glass. This layer increases in thickness and shows an interference color ranging from brown to blue when the following conditions are satisfied: optical thickness $nd = (2m + 1) \lambda/4$ (here m is an integer); when $m = 0$, $nd = \lambda_0/4$, $\lambda_0 \simeq 400$ to 600 nm. This is called staining.

2.2.3.3.   Formation Mechanism of Dimming

Dimming[18] refers to the following phenomenon: when a glass is allowed to stand in air, the glass comes into contact with the air, and a dew

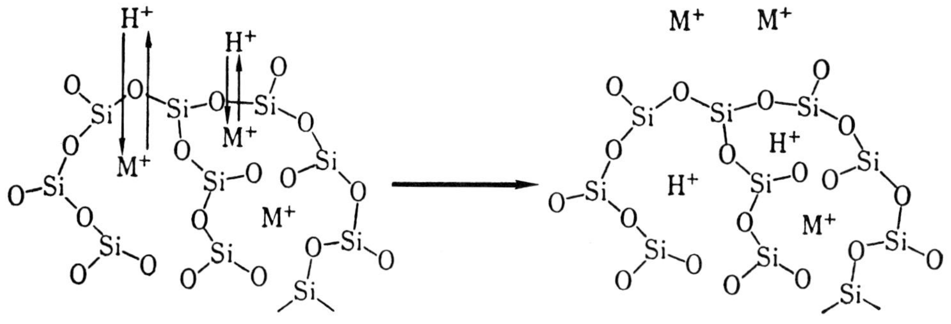

Figure 2.34. Reaction between glass and water.

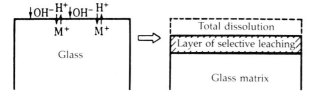

Figure 2.35. Reaction between glass surface and water.

2.2.3.4. Formation Mechanism of Latent Scratches

The causes of latent scratches lie in the polishing process.[19] When polishing is performed under harsher conditions (i.e., using a harder polisher, using a cerium oxide polishing agent fired at a higher temperature, and polishing under heavier pressure) latent scratches tend to occur. Immediately after polishing, the scratches are too small to be observable; when the glass is washed, however, the scratches are attacked by the detergent and become visible. In particular, minute scratches are enlarged by inorganic builders in detergents. Latent scratching shows a greater tendency to occur in glasses with a lower hardness or weaker chemical durability. In other words, latent scratches are minute scratches formed in the polishing process that are subsequently enlarged by chemical erosion in the washing process. Figure 2.38A and B shows the results of polishing SSK 4 using polytex and polyurethane as polishers. No scratches were generated in water or a neutral detergent when polytex was used, whereas scratches were generated in both water and a neutral detergent when the harder polyurethane was used.

Figure 2.39a and b shows the results of polishing using cerium oxide fired at 900°C and 1200°C. Numerous strong latent scratches were generated in case (b) where the harder polishing agent fired at 1200°C was used.

Figure 2.40a and b shows the results of a test in which the mechanical conditions of polishing, load, rpm, and mechanism of the polishing apparatus were varied. Specimen (a), which was polished under conditions bringing about approximately a four-fold increase in polishing speed, shows conspicuous latent scratches.

condenses on the glass (especially at night because of the difference in temperature between night and day); this dew attacks the surface of the glass. Because the mist consists of a small amount of water, the diffusion of alkali ions, etc., into the mist converts the dew into a concentrated alkali solution, which conversely erodes the silica gel layer. When the moisture evaporates as the temperature rises the following morning, silicate ions and alkali are deposited on the surface of the glass in the form of SiO_2, Na_2CO_3, $NaOH$, etc. This is the formation mechanism of dimming. The author et al., confirmed the products of dimming for several types of glasses using electron diffraction. Figure 2.36 shows the results. The deposited components were the same as the components dissolved when powdered glass samples were placed in platinum baskets and subjected to a water test using small amounts of water. This indicates that the products of dimming are formed by the deposition of the same components that are dissolved when the glass is subjected to a water-resistance test. Figure 2.37 shows percentages of components dissolved when powdered glass samples were subjected to water-resistance test. A special feature of the results is the fact that SiO_2 was deposited (unlike the case in an acid-resistance test).

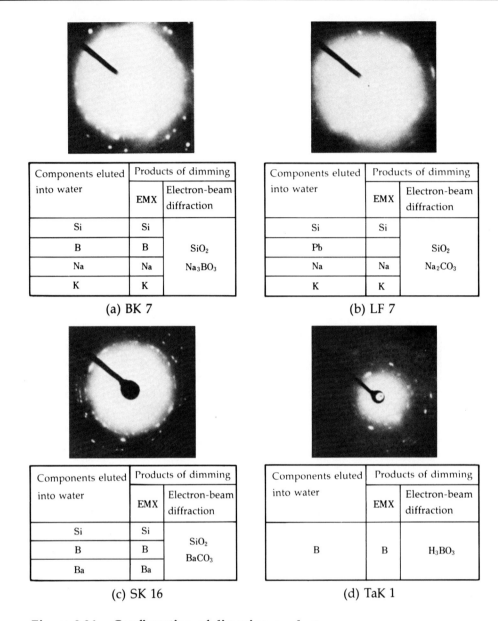

Components eluted	Products of dimming		Components eluted	Products of dimming	
into water	EMX	Electron-beam diffraction	into water	EMX	Electron-beam diffraction
Si	Si		Si	Si	
B	B	SiO₂	Pb		SiO₂
Na	Na	Na₃BO₃	Na	Na	Na₂CO₃
K	K		K	K	

(a) BK 7   (b) LF 7

Components eluted	Products of dimming		Components eluted	Products of dimming	
into water	EMX	Electron-beam diffraction	into water	EMX	Electron-beam diffraction
Si	Si	SiO₂			
B	B	BaCO₃	B	B	H₃BO₃
Ba	Ba				

(c) SK 16   (d) TaK 1

Figure 2.36.   Confirmation of dimming products.

When SSK 4 (H_k = 575 kg/mm²) and F 2 (H_k = 440 kg/mm²) which are both class 3 in terms of chemical durability but differ in terms of Knoop hardness (H_k), were compared, F 2 showed a greater tendency toward latent scratching. On the other hand, a comparison of SSK 4* and SK 16 (H_k = 570 kg/mm², class-5 acid resistance), which have approximately the same hardness but differ in terms of chemical durability, shows that SK 16 generates more latent scratches (Fig. 2.41).

_____

* SSK 4 is referred to in the original text, but Fig. 2.41 shows F 2.

2.2.3.5.   Methods of Evaluating the Chemical Durability of Optical Glasses

True Chemical Durability with Respect to Water (D_0). In order to evaluate the true chemical durability of an optical glass with respect to water, the evaluation method used must take into consideration the weight losses caused by both of the following: dissolution of the glass network broken by OH⁻ ions as a result of the reaction with water, and cation leaching resulting from ion exchange with H_3O^+ (Ref. 20). It is possible to measure the true chemical durability of a glass with respect to water (D_0) by immersing the polished

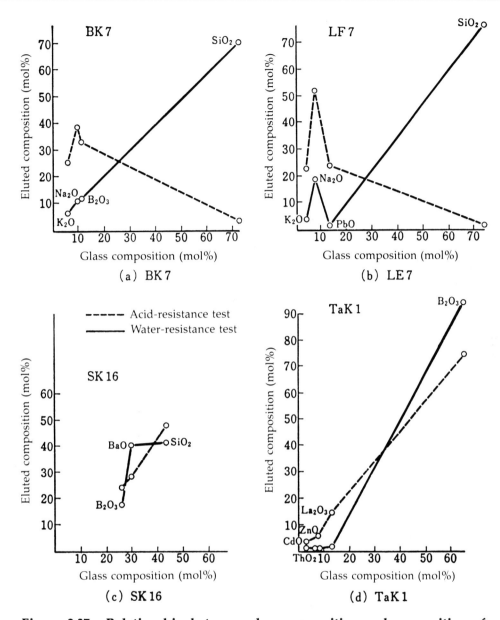

Figure 2.37. Relationship between glass composition and composition of eluted components.

glass in pure water circulating through an ion exchange membrane so that the two above mentioned reactions are allowed to proceed, and by measuring the weight loss that occurs at that time.

Figure 2.42 shows the test apparatus used. The samples were glass disks [43.7-mmϕ (30 cm^2), ~5-mm thick] with both sides polished. The edge surface of each disk was coated with a protective film of pitch. Each sample was immersed for a specified period of time in a 12-liter water bath at 50°C, with stirring at 1500 rpm. Afterward, the weight loss was measured, and the weight loss with respect to water per unit time and unit area D_0 (mg/cm$^2 \cdot$hr) was determined.

Resistance to Staining (T_{Blue}). The ease with which glass staining occurs in water is determined by direct measurement of the time required to produce staining because there are some glasses among optical glasses (e.g., phosphate glasses) in which a weight loss occurs but no interference color is exhibited. In Fig. 2.43, blank circles indicate F–SF systems, solid circles indicate BK–CF systems, blank diamonds indicate SK systems, blank triangles indicate LaCL–LaK systems, and solid triangles indicate NbF–TaF systems. As in the case of the samples used for D_0 measurement, polished glass disks were immersed in pure water that circulated through an ion exchange resin. The

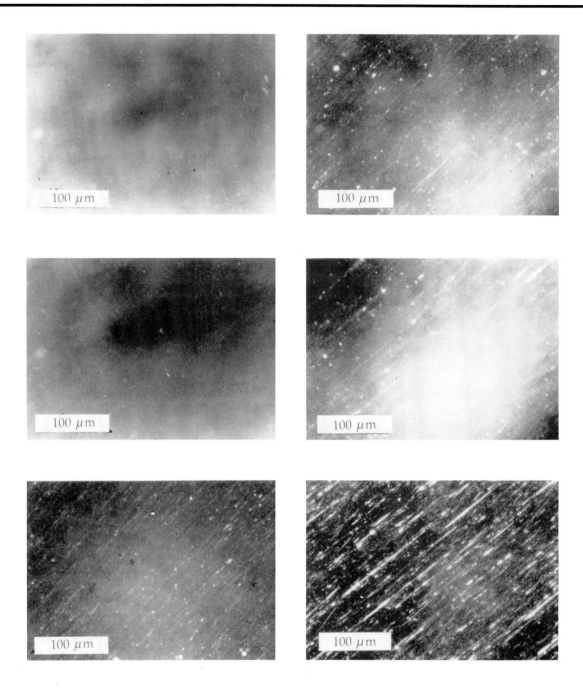

Figure 2.38A. Relationship between latent scratching and polisher. Polished surface of SSK 4 when polytex was used as a polisher. Washing conditions: (a) water, ultrasonic cleaning 120 s; (b) detergent, ultrasonic cleaning 40 s; (c) acid, ultrasonic cleaning 40 s.

Figure 2.38B. Relationship between latent scratching and polisher. Polished surface of SSK 4 when polyurethane was used as a polisher. Washing conditions: (a) water, ultrasonic cleaning 120 s; (b) detergent, ultrasonic cleaning 40 s; (c) acid, ultrasonic cleaning 40 s.

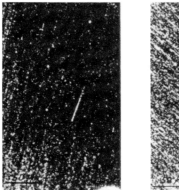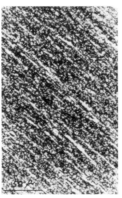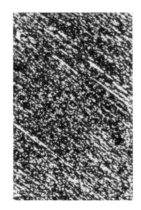

Figure 2.39. Relationship between polished surface and firing conditions of cerium oxide polishing agent: (a) fired at 900°C and (b) fired at 1200°C.

Figure 2.40. Effect of polishing conditions: (a) high-speed polishing (heavy pressure) and (b) low-speed polishing (light pressure).

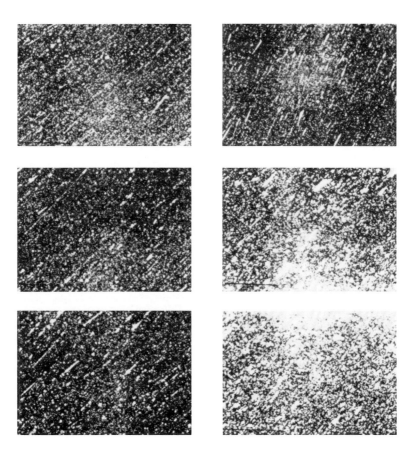

Figure 2.41A. Polished surface of F 2 and washing conditions: (a) water, ultrasonic cleaning 120 s; (b) detergent, ultrasonic cleaning 10 s; (c) acid, ultrasonic cleaning 10 s.

Figure 2.41B. Polished surface of SK 16 and washing conditions: (a) water, ultrasonic cleaning 120 s; (b) detergent, ultrasonic cleaning 40 s; (c) acid, ultrasonic cleaning 40 s.

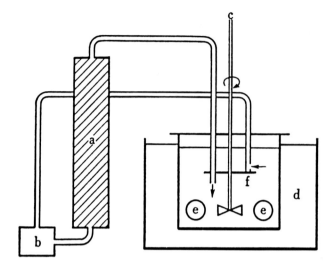

a. Ion-exchange resin
b. Circulating pump
c. Stirrer
d. Constant-temperature water vat
e. Glass samples
f. Pure water vat

Figure 2.42. Apparatus used to measure durability.

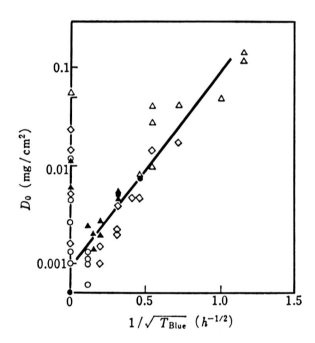

Figure 2.43. True chemical durability with respect to water and apparent rate of staining formation in water.

time (T_{Blue}) required for the interference color of the staining layer thus generated to change from brown to blue ($nd \simeq 120$ to 130 nm) was measured. The rate of formation of staining can be evaluated by $1/\sqrt{T_{Blue}}$. As shown in Fig. 2.43, there are some glasses (SK, PSK) in which the weight loss occurs almost exclusively as a result of dissolution of the network so that no staining occurs,

and other glasses (SK, LaCL) in which the weight loss occurs as a result of both dissolution of the network and ion exchange of alkali earths so that staining tends to occur. The BK, F, and TaF glasses show little weight loss and little tendency for staining to occur.

Kanamori et al.[21] measured the indices of refraction and thicknesses of staining layers generated for silicate glasses, borate glasses, and phosphate glasses through the measurement of spectroscopic reflectance. They also clarified the structure of the staining layer. The index of refraction and thickness of the staining layer can be determined from the following equations[22]:

$$[R_f] = 1$$

$$- \frac{4 n_{fa} n_{fg} n_g}{(n_{fa} n_g + n_{fg})^2 - (n_{fa}^2 - 1)(n_g^2 - n_{fg}^2) \sin^2 1/2 \delta_0} \ ,$$

$$(2.14)$$

$$\delta_0 = \frac{4\pi}{\lambda} \int_0^{d_f} n(z) dz \approx \frac{2\pi}{\lambda} (n_{fa} + n_{fg}) d_f \ . \qquad (2.15)$$

Here, the maximum and minimum indices of refraction R_{max} and R_{min} can be expressed as follows:

$$(n_{fa} + n_{fg}) d_f = 2m\lambda/2, R_{max} = \frac{(n_{fa} n_g - n_{fg})^2}{(n_{fa} n_g - n_{fg})^2} \ ,$$

$$(2.16)$$

$$\left(n_{fa} + n_{fg}\right) d_f = (2m + 1)\lambda/2, \quad R_{min} = \frac{\left(n_{fa}n_{fg} - n_g\right)^2}{\left(n_{fa}n_{fg} + n_g\right)^2}.$$

$$(2.17)$$

Here, n_g is the index of refraction of the glass, n_{fg} is the index of refraction of the glass side of the staining layer, n_{fa} is the index of the refraction of the atmosphere side of the staining layer, n_0 is the index of refraction of the atmosphere, and d_f is the thickness of the staining layer. The respective spectroscopic reflectance curves a, b, and c in Fig. 2.44 correspond to type a, b, and c staining layers in Fig. 2.45. The values n_{fg}, n_{fa}, and d_f can be determined by inserting measured values of R_{max}, R_{min}, λ_{max}, λ_{min}, and n_g into Eqs. (2.16) and (2.17).

The Staining Layer in Borate Glasses. The compositions of the glasses used are shown in Table 2.10A; n_{fg}, n_{fa}, and d_f of the staining layers obtained and the water-resistance weight losses are shown in Table 2.10B. The B_2O_3–BaO glasses that contain little La_2O_3 show a high D_0; the staining layer is fragile and peels easily. When Ba^{2+} is replaced by La^{3+}, or by Mg^{2+}, Ca^{2+}, or Sr^{2+}, an interference color is observed, and n_{fg}, n_{fa}, and d_f can be determined. The model shown in Fig. 2.46 was inferred for the staining layer formed by the erosion of borate glasses in water; this model was inferred from the following values: $n_{fg} = 1.41$, $n_{fa} = 1.36$, $d_f = 270$ to 970 nm, index of refraction of the glass $n_g = 1.63$ to 1.67. In the case of borate glasses, B_2O_3 and BaO are soluble in water; accordingly, the structural components of the staining layer would appear to be La oxide and hydroxide components. The index of refraction of the staining layer is approximately 1.40 and is thus considerably lower than that of bulk glass ($n_g \sim 1.65$); this indicates that the structure is extremely porous.

The Staining Layer in Silicate Glasses. The compositions of the silicate glasses used are shown in Table 2.11A; n_{fg}, n_{fa}, and d_f of the staining layers obtained, and the water-resistance weight losses D_0 are shown in Table 2.11B.

In Na_2O–SiO_2 binary glasses, an extremely thick staining layer (several microns thick) is formed. When Al_2O_3 is introduced into this system, the chemical durability is greatly improved so that no staining layer is formed. In the case of system containing Li^+, Mg^{2+}, Ca^{2+}, and Sr^{2+}, an interference color is observed; the staining layer increases in thickness with immersion time, as does the irregularity of the surface of the staining layer. The distribution of the index of refraction of

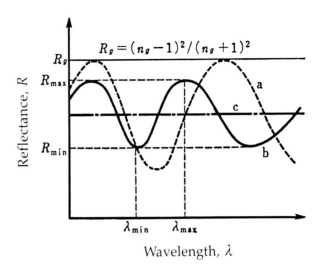

Figure 2.44. Spectroscopic reflectance curves of the staining layer.

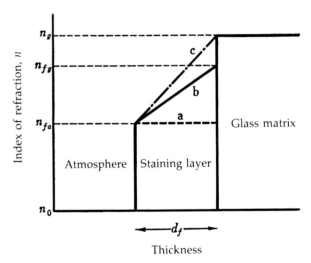

Figure 2.45. Model of refractive index distribution of the staining layer.

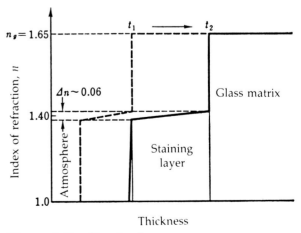

Figure 2.46. Borate glasses.

Table 2.10A.   Composition of borate glasses (mol%)

	B₂O₃/BaO					La₂O₃/BaO			RO				B₂O₃/SiO₂		
	B2	1	3	4	5	B1	6	7	B8	9	10	11	B12	13	14
B_2O_3	80	75	70	65	60		75			57.5			70	62.5	55
SiO_2			—				—			—			5	12.5	20
La_2O_3			5			5	10	15		7.5				5	
RO	15	20	25	30	35	20	15	10		35.0				20	
			Ba				Ba		Mg	Ca	Sr	Ba		Ba	

Table 2.10B.   Properties of the staining layer in borate glasses.

	Sample glass		D_0 (10⁻³ mg/cm²·h)	n_g	$n_{fg}{}^a$	$n_{fa}{}^a$	$d_f{}^a$ (nm)
B_2O_3/BaO ratio	H ↑ ↓ L	B2	300	1.60	?	?	?
		1	255	1.62	?	?	?
		3	270	1.64	?	?	?
		4	345	1.66	?	?	?
		5	620	1.67	?	?	?
BaO/La_2O_3 ratio	H ↑ ↓ L	B1	255	1.62	?	?	?
		6	28	1.66	1.41	1.37	430
		7	19	1.69	?	?	–
B_2O_3/SiO_2 ratio	H ↑ ↓ L	B12	275	1.63	?	?	?
		13	170	1.63	?	?	?
		14	75	1.63	1.52	1.41	970
RO	Mg	B8	16	1.64	1.38	1.35	270
	Ca	9	25	1.68	1.41	1.36	340
	Sr	10	83	1.67	1.42	1.36	970
	Ba	11	410	1.69	?	?	?
BaCD 16			26	1.62	?	?	<70

ᵃ Immersion time: 5 hours.

the staining layer can be expressed by the model shown in Fig. 2.47. In the initial stages of erosion, the staining layer is formed nearly homogeneously; a slight gradient in index of refraction begins to be seen as the reaction proceeds. Compared with the staining layer seen in borate glasses, the staining layer in silicate glasses is a much stronger and harder film.

Staining in Phosphate Glasses. The compositions of the phosphate glasses used and the results of erosion by pure water are shown in Table 2.12A and B.

In phosphate glasses, as shown in Table 2.11B, the glass as a whole is dissolved so that no staining layer is formed. Accordingly, a model diagram indicating the distribution of the index of refraction of the immersion-treated glass surface can be drawn as shown in Fig. 2.48.

Resistance to Dimming (D_w). One method of measuring dimming is to artificially induce dimming, and then measure the degree of white cloudiness with a haze meter. However, the reproducibility of mist droplet formation is poor, and the inadequate precision of measurement makes a detailed classification impossible. The formation mechanism of dimming consists of four stages:

1.   Adhesion of dew droplets.
2.   Dissolution of soluble ions into the dew droplets and alkalinization of the dew droplets.
3.   Erosion of the silica layer by the resulting alkaline solution.
4.   Deposition of the dissolved components.

In the powder water-resistance weight-loss method, powdered glass samples are packed into platinum baskets and immersed in small amounts

Table 2.11A. Silicate glasses.

	SiO$_2$/Na$_2$O			Al$_2$O$_3$/BaO			R$_2$O			RO			
	S1	2	10	S3	4	11	S5	3	6	S7	8	9	3
SiO$_2$	65	75	85		65			65			65		
Al$_2$O$_3$		—		0	5	10		—			—		
R$_2$O	35	25	15		15			15			15		
		Na			Na		Li	Na	K		Na		
RO		—		20	15	10		20			20		
					Ba			Ba		Mg	Ca	Sr	Ba

Table 2.11B. Properties of the staining layer in silicate glasses.

Sample glass			D_0 (10^{-3} mg/cm^2)	n_g	n_{fg}[a]	n_{fa}[a]	d_f[a] (nm)
Na$_2$O/SiO$_2$ ratio	H ↕ L	S1	12000				
		2	77	1.50	(3) 1.50	1.26	~14 μm
		10	1.6	1.48	(24) 1.48	1.47	~ 6 μm
					(45) 1.48	1.46	~ 7 μm
BaO/Al$_2$O$_3$ ratio	H ↕ L	S3	160	1.57	(5) ?	?	?
		4	(0)	1.56			0
		11	(0)	1.54			0
R$_2$O	Li	S5	4.0	1.59	(10) 1.48	1.48	330
					(24) 1.48	1.48	570
					(45) 1.49	1.47	760
					(72) 1.49	1.47	850
	Na	3	160	1.57	(5) ?	?	?
	K	6	420	1.57	(5) ?	?	?
RO	Mg	S7	0.4	1.52	(72) ?	?	<50
	Ca	8	1.3	1.55	(24) ?	?	~70
	Sr	9	5.6	1.55	(10) 1.50	1.47	610
					(24) 1.51	1.47	870
					(45) 1.51	1.46	1060
					(72) 1.51	1.44	1240
	Ba	3	160	1.57	?	?	?

[a] Immersion time in hours.

of water, and the weight loss is measured. This method consists of the following steps:

1. Immersion of a powdered-glass sample in a small amount of water.

2. Dissolution of soluble ions into the small amount of water, and alkalinization of the test solution by the dissolved ions.

3. Erosion of the SiO$_2$-rich layer by the alkaline solution.

Thus, except for the deposition of the dissolved components, this method corresponds to the formation mechanism of dimming. Because all of the components dissolved into the dew droplets are deposited in dimming, all of the dissolved components are measured as a weight loss in the powder weight-loss method. Accordingly, we may consider a correspondence to exist between the rate of dimming formation and water-resistance weight loss in the powder method; by measuring the weight loss, it is possible to ascertain indirectly the tendency for dimming to occur.

Figure 2.49 shows the relationship between rate of dimming formation Dw·ρ and D_0. The following four categories of glasses are apparent:

1. Glasses in which the weight loss results almost exclusively from dissolution of the network so that dimming does not readily occur (LaK).

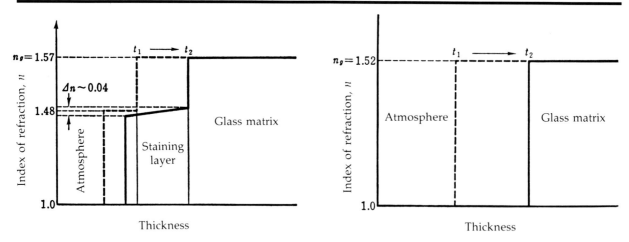

Figure 2.47.   Silicate glasses.   **Figure 2.48.   Phosphate glasses.**

Table 2.12A.   Phosphate glasses.

	P_2O_5/Na_2O			Al_2O_3/CaO			R_2O			RO			
	P2	1	3	P9	10	11	P4	1	5	P6	1	7	8
P_2O_3	50	55	65		60			55			55		
Al_2O_3		5		5	10	15		5			5		
R_2O	25	20	10		20			20			20		
		Na			Na		Li	Na	K		Na		
RO		20		15	10	5		20			20		
		Ca			Ca			Ca		Mg	Ca	Sr	Ba

Table 2.12B.   Properties of the staining layer in phosphate glasses.

	Sample glass		D_0 (10^{-3} mg/cm^2)	n_g	n_f[a] (10 hr)[b]	n_f[a] (48 hr)[b]	d_f
Na_2O/P_2O_5 ratio	H	P2	11.6	1.52	1.52	1.52	0
	↕	1	4.5	1.52	1.52	1.52	0
	L	3	4.8	1.52	1.52	1.52	0
CaO/Al_2O_3 ratio	H	P9	7.3	1.51	1.51	1.51	0
	↕	10	0.3	1.52	1.51	1.51	0
	L	11	0.3	1.52	1.52	1.52	0
R_2O	Li	P4	3.9	1.53	1.53	1.53	0
	Na	1	4.5	1.52	1.52	1.52	0
	K	5	10.7	1.51	1.51	1.51	0
RO	Mg	P6	1.9	1.51	1.51	1.51	0
	Ca	1	4.5	1.52	1.52	1.52	0
	Sr	7	2.2	1.52	1.52	1.52	0
	Ba	8	2.7	1.54	1.53	1.53	0

[a] $n_f = n_{fg} = n_{fa}$.
[b] Immersion time.

2.   Glasses in which the weight loss involves mainly the leaching of divalent cations and dissolution of the network so that dimming (and staining) readily occurs (LaCL, SK).

3.   Glasses in which the leaching of alkali and Pb^{2+} ions takes place, but in which there is little elution of the network (SF).

4.   Glasses in which there is almost no dissolution of the glass network, with only alkali being leached (BK, KF, F). A special feature of this system is that dimming readily occurs even though there is no weight loss.

Optical glasses, therefore, can be roughly divided into five characteristic types:

1.   Glasses in which both ion exchange and network dissolution occur (SK, LaCL, etc., where both staining and dimming occur).

2.   Glasses in which there is almost no occurrence of either of the aforementioned phenomena (TaC, TaF).

3.   Glasses in which only ion leaching occurs (BK, FL, CF, etc., where almost no staining occurs, but dimming does occur).

4.   Glasses in which only B_2O_3 is leached (NbF, NbSF, where some staining, but no dimming occurs).

5.   Glasses in which the network as a whole is dissolved (PKS, where no staining occurs).

Resistance to Detergents (D_{NaOH}, D_{STPP}). In the phenomenon of latent scratching, minute scratches that are created during the polishing process are enlarged by inorganic builders in the detergent during the washing process. The ease with which scratching occurs depends on the hardness of the glass, whereas the tendency of the scratches to grow depends on the chemical durability of the glass with respect to the inorganic builders involved. Representative examples of inorganic builders are Na_2CO_3 and $Na_5P_3O_{10}$. Accordingly, the resistance can be evaluated by means of the weight loss that occurs when samples are immersed in 1/100 M solutions of these substances. As seen in Fig. 2.50, SK glasses are attacked by NaOH, whereas Pb and La glasses tend to dissolve in $Na_5P_3O_{10}$. This behavior is

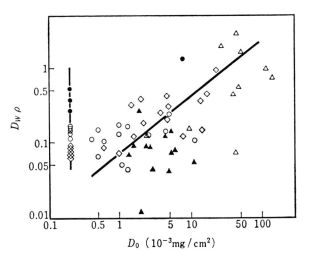

Figure 2.49.   True chemical durability with respect to water and resistance to dimming.

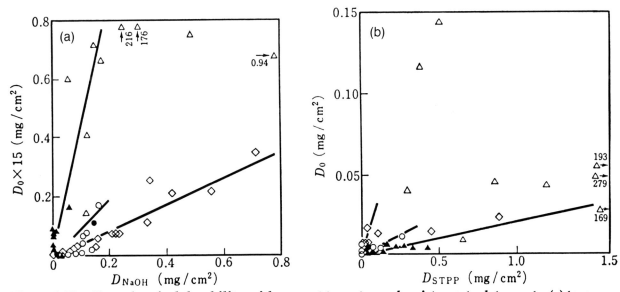

Figure 2.50.   True chemical durability with respect to water and resistance to detergents: (a) true chemical durability with respect to water and detergent resistance, D_{NaOH}; and (b) true chemical durability with respect to water and detergent resistance, D_{STPP}.

thought to result from the dissolution of large amounts of Pb and La in the $Na_5P_3O_{10}$ solution in the form of $[PbP_3O_{10}]^{-3}$ and $[LaP_3O_{10}]^{-2}$.

Table 2.13 shows values of D_0, T_{Blue}, D_w, D_{NaOH}, and D_{STPP} for 70 of the most highly demanded types of optical glasses.

Table 2.13. Chemical durability of optical glasses.

Glass	D_0	T_{Blue}	D_w	D_{NaOH}	D_{STPP}	Glass	D_0	T_{Blue}	D_w	D_{NaOH}	D_{STPP}
FC 5	7.5	5.0	0.54	0.15	0.02	BaF 3	<0.3	>70	0.02	0.05	<0.01
BSC 7	<0.3	>70	0.21	0.06	<0.01	BaF 10 N	<0.3	>70	0.02	0.03	<0.01
C 3	<0.3	>70	0.15	0.04	<0.01	BaF 11 N	<0.3	>70	0.02	0.02	<0.01
CF 6 N	<0.3	>70	0.06	0.02	<0.01	BaF 13	5.0	–	0.08	0.22	0.05
FEL 2	<0.3	>70	0.09	0.04	<0.01	BaF 22 N	0.4	a	0.02	0.02	<0.01
FEL 6	<0.3	>70	0.05	0.03	<0.01	BaFD 2	1.6	a	0.03	0.07	0.04
FL 5	<0.3	>70	0.04	0.05	<0.01	BaFD 7 A	<0.3	>70	0.02	<0.01	<0.01
FL 7	<0.3	>70	0.04	0.05	<0.01	BaFD 8 N	<0.3	>70	0.02	<0.01	<0.01
F 1	0.4	>70	0.03	0.08	<0.01	BaFD 15	<0.3	>70	0.03	<0.01	<0.01
F 2	0.6	70	0.03	0.06	<0.01	LaCl 2	45.0	2.0	0.80	0.94	0.85
F 4	<0.3	>70	0.04	0.05	<0.01	LaCl 5	40.0	3.5	0.11	0.06	0.29
F 5	<0.3	>70	0.04	0.05	<0.01	LaCl 6	50.5	a	0.44	0.49	0.93
FDS 3	5.0	a	0.05	0.12	0.07	LaCl 7	27.0	3.5	0.56	0.12	0.69
FD 1	2.6	a	0.03	0.11	0.04	LaC 8	5.3	a	0.02	<0.01	0.27
FD 2	1.1	70	0.01	0.07	0.02	LaC 9 N	11.0	a	0.02	0.03	0.65
FD 4	4.5	a	0.03	0.11	0.12	LaC 10	4.5	10	0.03	<0.01	0.42
FD 5	1.0	70	0.04	0.09	0.02	LaC 11	42.5	2.0	0.02	0.17	0.19
FD 6	11.2		0.02	0.16	0.26	LaC 12	47.8	1.0	0.14	0.15	0.79
FD 3	1.3	70	0.04	0.08	0.04	LaC 13 N	5.4	10	0.04	0.01	0.22
FD 10	0.3	>70	0.03	0.08	<0.01	LaC 14 A	5.1	a	0.01	0.01	0.15
FD 11 N	1.3	a	0.01	0.14	0.03	TaC 6	3.3	70	0.01	<0.01	0.03
FD 13	1.0	a	0.03	0.08	0.02	LaFL 2	116.0	0.8	0.25	0.31	0.38
FD 14	0.5	a	0.01	0.10	<0.01	LaFL 3 N	4.6	45	0.03	0.02	0.01
FD 15 N	0.5	a	0.04	0.08	<0.01	LaFL 5 N	2.0	45	0.01	<0.01	0.01
BaCl	2.2	10	0.06	0.08	<0.01	LaFL 6 N	0.9	72	0.01	<0.01	<0.01
BaC 4 N	0.3	>70	0.02	0.03	<0.01	LaF 2	2.5	25	0.03	<0.01	0.02
BaCD 2	1.5	25	0.09	0.09	0.02	LaF 3 N	5.7	45	0.04	<0.01	0.04
BaCD 4	4.7	6.0	0.06	0.21	0.01	NbF 1	5.0	10	0.01	<0.01	0.33
BaCD 5	2.0	10	0.12	0.15	<0.01	NbFD 3	1.9	25	0.06	<0.01	0.14
BaCD 7	4.7	5.0	0.12	0.24	<0.01	NbFD 10	1.6	25	0.02	<0.01	0.03
BaCD 10	7.8	5.0	0.03	0.33	0.04	NbFD 11	2.7	25	0.02	<0.01	0.12
BaCD 14	17.0	2.0	0.13	0.34	0.04	NbFD 12	1.9	25	0.01	<0.01	0.07
BaCD 15 N	11.0	10	0.09	0.07	0.05	NbFD 13 N	1.4	45	0.02	<0.01	0.02
BaCD 16 N	23.4	a	0.07	0.71	0.89	TaF 1 A	2.4	70	0.01	<0.01	0.02
BaCD 18 N	9.9	10	0.03	0.08	0.05	TaF 3 A	1.2	70	0.01	<0.01	0.02
BaCED 4	3.8	10	0.07	0.16	0.02	TaFD 1 N	2.5	70	0.01	<0.01	<0.01
BaCED 5 N	1.0	25	0.02	0.03	<0.01	TaFD 5	0.4	70	0.01	<0.01	0.02

a Glass types in which no blue staining layer is observed because of dissolution of the glass surface as a whole, or glass types in which the change of the blue interference color is irregular.

References

1. T. Izumitani and K. Nakagawa, *VIIth International Glass Conference, Brussels, Belg.* **5**, 1 (1965).
2. E. Kordes, *Glastech. Ber.* **17**, 65 (1939).
3. K. Fajans and N. J. Kreidl, *J. Am. Ceram. Soc.* **31**, 105 (1948).
4. T. Isumitani, *Bulletin, Osaka Ind. Res. Inst.* **7**, 1, 35 (1956).
5. T. Izumitani, Terai, and Hamamura, *Bulletin Osaka Ind. Res. Inst.* **6**, 4, 46 (1955).
6. T. Izumitani and S. Hirota, *J. Non-Cryst. Solids* **29**, 109 (1978).
7. T. Izumitani, H. Sagara, H. Toratani, and S. Hirota, *Beiging International Symposium on Glass* **8**, 25 (1981); *J. Non-Cryst. Solids*, Special issue (in press).
8. T. Izumitani and S. Hirota, *IInd Internationales Otto-Schott-Kolloquium* (Jena, East Germany, 1982).
9. H. R. Philipp, *Solid State Commun.* **4**, 73 (1966).
10. G. H. Sigel, Jr., *J. Phys. Chem. Solids* **32**, 2373 (1971).
11. A. R. Ruffa, *Phys. Status Solids* **29**, 605 (1968).
12. A. J. Bourdillon, F. Khumalo, and J. Bordas, *Phil. Mag. B* **37**,731 (1978).
13. F. Seitz, *J. Chem. Phys.* **6**, 150 (1938).
14. M. Imaoka, *Tokyo Daigaku Seisun Gijutsu Kenkyusho Hokoku* (Tokyo University Technical Publication Research Report) **6**, 4, 127 (1957); *Yogyo-Kyokai-Shi* **69**, 282 (1961).
15. T. Izumitani and K. Nakagawa, *Phys. Chem. Glasses*, 13, **3**, 85 (1972).
16. I. Prigogine, *Bull. Soc. Chim., Belg.* **52**, 15 (1943).
17. Tsuehi Hashi, *Physical Chemistry of Glass Surfaces [Garasu Hyōmen no Batsuri Kagaku]*, Kodansha Ltd., Tokyo (1980).
18. T. Izumitani and E. Miyade, *Xth International Glass Conference, Kyoto* **9**, 55 (1974).
19. T. Izumitani and E. Miyade, S. Adachi, and S. Harada, *National Bureau of Standards Special Publication 562*, 417 (1979).
20. T. Izumitani, E. Miyade, and S. Adachi, *J. Non-Cryst. Solids* **42**, 1 to 3, 569 (1980).
21. T. Izumitani and C. Kanamori, *Hoya Private Seminar*, Cologne (1982).
22. K. Kinoshita, *Progress in Optics IV* (North Holland and Publishing Company, Amsterdam, 1965), 115.

CHAPTER 3

How Optical Glasses are Produced: Manufacturing Methods and Progress in Manufacturing Methods

3.1. Fundamentals of Optical Glass Manufacturing Processes and Manufacturing Techniques

The optical glass manufacturing process consists of three processes: melting, forming, and annealing. As described in Chapter 2, a point of difference between optical glasses and other glasses is that optical glasses are homogeneous; in simpler terms, no striae are present in such glasses. Furthermore, this means that such glasses contain few bubbles. It is no exaggeration to say that optical glass technicians direct their total efforts toward melting techniques to eliminate bubbles and striae. Defects generated in the melting process are devitrification, striae, and bubbles. Devitrification depends on the composition of the glass (discussed in Chapter 2). Optical glass manufacturing technology concentrates on ways of eliminating striae and bubbles during melting.

The manufacturing process continues from melting to pressing. In the past, hand pressing was performed: recently, however, direct pressing in which the molten glass is immediately pressed has come into use. Pressed lens blanks are precision annealed, residual stain is eliminated, and the index of refraction is adjusted to the desired value. In this chapter, we will discuss the fundamentals of these various techniques.

3.1.1. Melting

3.1.1.1. Striae

Striae and Their Causes. Striae are regions of the glass that have a different index of refraction from the homogeneous matrix. Because these regions usually exhibit a thread-like appearance, they are called striae. This difference in the index of refraction arises from a difference in chemical compositions, it is not caused by physical factors. Figure 3.1 shows a photograph of striae projected by means of a zircon lamp. Striae are observed by

the projection methods, the schilleren method, the interference method, or visually.

Figure 3.2 shows specific gravity distribution curves for striae and homogeneous regions obtained by centrifugal separation; the difference

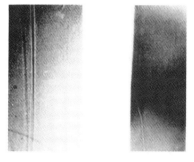

Figure 3.1 Striae projected by means of a zircon lamp.

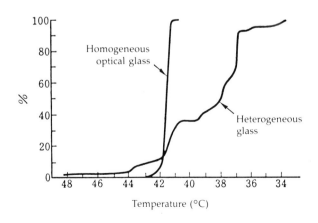

Figure 3.2. Specific gravity distribution curves for homogeneous optical glass and heterogeneous borosilicate glass.

between homogeneous optical glass and heterogeneous borosilicate glass can be clearly seen. Here, the abscissa indicates the temperature of the heavy liquid used for specific gravity separation. The abscissa, therefore, corresponds to the specific gravity of the suspended glass particles, whereas the ordinate indicates the percentage of particles reaching the top end of the tube.

Figure 3.3 shows specific gravity distribution curves of striae in BK 7 glasses that were melted using clay pots. These striae all consist of low-specific-gravity portions; therefore, these striae contain greater amounts of Al_2O_3 and SiO_2 as a result of erosion of the clay pots. In addition, as

shown in Fig. 3.4, the striae in lanthanum glasses consist of high-specific-gravity portions, indicating that these striae were produced by the volatilization of B_2O_3. Thus, striae are created when heterogeneous portions, generated by the erosion of refractories or the volatilization of glass components, are drawn into the glass.

What kind of difference in index of refraction exists between striae and homogeneous regions? Examples of measurements performed for fairly large and clear striae (using an interferometer) are shown in Table 3.1. Striae show indices of refraction that are 1×10^{-4} to 5×10^{-5} higher or lower than those of homogeneous regions. Furthermore,

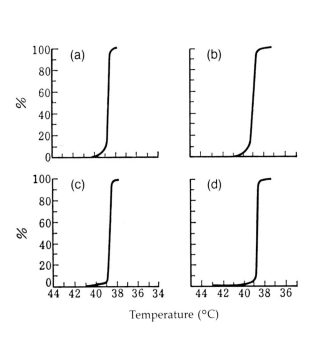

Figure 3.3.   Specific gravity distribution curves of striae (BK 7).

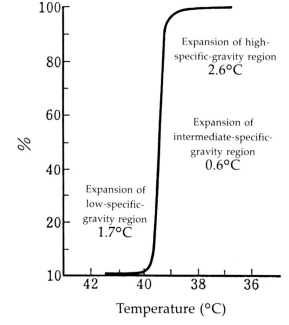

Figure 3.4.   Specific gravity distribution curves of striae (LaF 4).

Table 3.1.   Differences in index of refraction shown by striae in various glasses (a is the thickness of the striae).

		A[a]	B[a]	C[a]	D[a]
BK 7	a	2.12 mm	3.86		1.66
	Δn	-5.2×10^{-5}	-7.1×10^{-5}		-1.6×10^{-4}
F 2	a	1.64 mm	2.50	2.44	4.59
	Δn	-1.0×10^{-4}	-1.1×10^{-4}	-4.5×10^{-5}	-3.0×10^{-5}
F 16	a	2.26 mm	3.33	2.44	3.97
	Δn	$+3.6 \times 10^{-5}$	$+4.9 \times 10^{-5}$	$+3.4 \times 10^{-5}$	$+2.8 \times 10^{-5}$
LaK 12	a	0.64 mm	1.66		
	Δn	$+1.7 \times 10^{-4}$	$+5.8 \times 10^{-5}$		

[a] Lettered columns refer to volumes within the melt as shown in Fig. 3.5.

it can be inferred from Table 3.1 that striae in BK 7 and F 2 are generated by pot erosion, striae in F 16 by the volatilization of fluorides, and striae in LaK 12 by the volatilization of B_2O_3.

Striae and Surface Tension. Striae can be removed by using forced stirring to cause diffusion of the striae portions; factors that affect the disappearance of striae by diffusion are the viscosity and surface tension of the melt. As the viscosity decreases, striae portions tend to disappear through diffusion with greater readiness; as the surface tension increases, they disappear less readily.

Striae in optical glasses are created by pot erosion or surface volatilization. Striae created by pot erosion contain large amounts of SiO_2 and Al_2O_3, and thus have a high surface tension. Furthermore, when fluorides or B_2O_3 are volatilized from the surface, the surface tension increases.[1] Moreover, because the surface tension of striae in optical glasses is higher than the surface tension of homogeneous portions, we may conclude that such striae would, in general, not readily disappear. Jebsen-Marwedel[2] reports that homogenization occurs as a result of homogeneous liquid being drawn into the striae, or as a result of striae being drawn into the homogeneous liquid, when the following condition for dynactivity is satisfied:

$$\left(\sigma_1 - \sigma_2\right)\left(\rho_1 - \rho_2\right) < 0 \quad .$$

[Reviewer's note: assume σ is surface tension and ρ is specific gravity.] However, whereas striae that are rich in SiO_2 and Al_2O_3 have a low specific gravity and a high surface tension so that mixing by dynactivity will occur; striae created by the volatilization of fluorides or B_2O_3 have a high specific gravity and a high surface tension so that mixing by dynactivity cannot be expected. In the latter case, therefore, it appears that strong stirring is required to eliminate striae.

Stirring. A point of difference between optical glass manufacturing methods and other glass manufacturing methods is that stirring is used in the manufacture of optical glasses. This is something that has been true since the beginning of optical glass manufacture, and various types of stirring have been tried. Platinum pots came into use for the melting of lanthanum glasses in 1945, before that time clay pots were used as melting pots for optical glasses. The use of clay pots resulted in erosion of the pot by the glass during melting, and this was unavoidable. Accordingly,

the main reason for stirring was not to diffuse the striae, but was to create a constant current in the pot by stirring so that the striae were fixed in the wall areas and central area of the pot (with only the homogeneous portions surrounded by these areas being extracted). Figure 3.5 shows this concept. The use of platinum pots made the consideration of striae created by refractory erosion unnecessary; as a result, it became possible to select stirring rods with good stirring efficiency for the purpose of eliminating striae created by volatilization.

Figure 3.6 shows the flow configuration that occurs when a conventional stirring rod is used. Colored glycerin poured into the center of the surface is pushed around the periphery by the motion of the rod and moves along the peripheral walls to the bottom; in the central area, the glycerin remains as striae. This configuration of residual striae corresponds to the configuration of residual striae in actual clay pot melting.

The author et al.,[3] investigated the stirring efficiency of the rod-type, crank-type, propeller-type, vertical-vane-type, and screw-type stirring rods shown in Fig. 3.7. The investigations showed that crank-type stirring rods have a high stirring efficiency at low viscosities, whereas screw-type stirring rods have a high stirring efficiency at high viscosities. Figure 3.8 shows the liquid-flow configurations that occur when crank-type and screw-type stirring rods are used. When stirring is

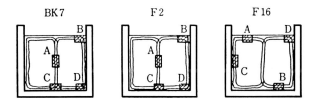

Figure 3.5. Examples of the distribution of striae in clay pots.

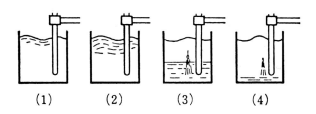

Figure 3.6. Liquid flow caused by a conventional rod-type stirring rod (60 rpm).

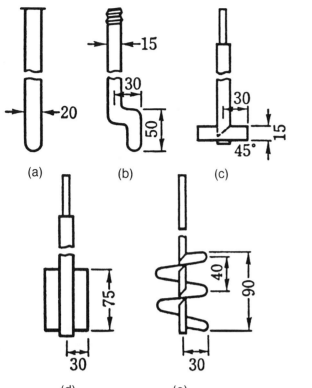

Figure 3.7. Shapes and dimensions of stirring rods (mm).

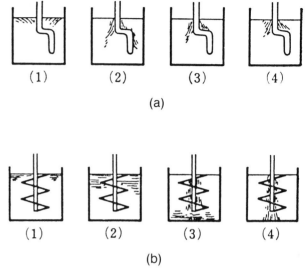

Figure 3.8. Liquid flow caused by crank-type and screw-type stirring rods (60 rpm).

performed using a crank-type stirring rod, colored glycerin is drawn into the interior area along the shaft. At the bend in the rod, the glycerin is expelled toward the periphery, and disappears. Striae are forcibly diffused all the way to the bottom of the pot so that no standing striae are generated in the central area. On the other hand, there is a tendency for striae to remain in the vicinity of the shaft near the surface, but if stirring is performed at high rpm, there are almost no residual striae. In the case of the screw-type stirring rod, colored glycerin is drawn down from the surface along the peripheral walls; glycerin that has reached the bottom is gradually picked up by the screw-form vane and is diffused into the solution as a whole as it circulates around the stirring rod. A special feature of this process is that there is not much change in this mixing mechanism even if the viscosity of the liquid varies. Such crank-type stirring rods have shown great effectiveness in the stirring of lanthanum glasses. The liquid flow configuration produced by such a stirring rod, and the fact that high-speed revolution is possible without entraining bubbles, result in the elimination of central striae. Consequently, such stirring rods

have made a great increase in yield possible (i.e., an increase from 60 to 95%).

The following relationship exists between the rotational speed of the stirring rod N and the time required for completion of mixing θ:

$$N^a \cdot \theta = k \quad . \tag{3.1}$$

Figures 3.9 through 3.11 show the $N-\theta$ relationship for dilute glycerin ($\eta = 0.15 \times 10^1$ poise), concentrated glycerin ($\eta = 1 \times 10^1$ poise), and a concentrated solution of millet jelly ($\eta = 2 \times 10^2$ poise). The relationship in Eq. (3.1) holds true in all three cases. A dilute glycerin solution corresponds to a molten glass viscosity of 4 to 5 poise and thus corresponds to the viscosity of lanthanum glasses. In this region, the crank-type stirring rod is the most effective at high speeds. At low speeds, however, an unmixed portion remains in the vicinity of the surface so that the time required for completion of mixing is increased. The screw-type stirring rod shows a gradual slope. A concentrated glycerin solution corresponds to glasses with a viscosity of about 100 poise: lanthanum glasses containing large amounts of SiO_2 would be included in this category. Here, the crank-type stirring rod again produces effective mixing at high speeds, whereas the screw-type stirring rod produces effective mixing at low speeds. The values of a and k increase considerably with an increase in viscosity. A dilute millet jelly solution corresponds to a molten glass viscosity of 150 to

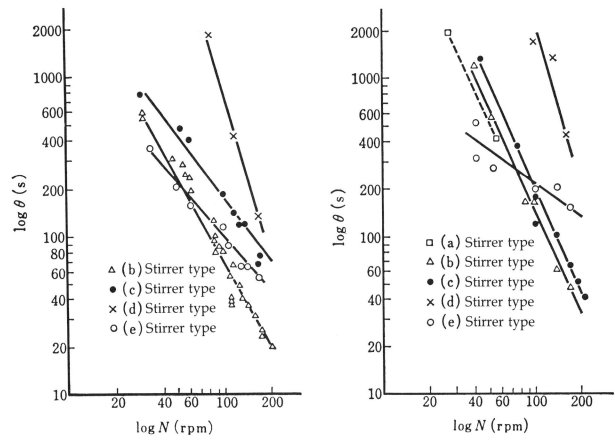

Figure 3.9. The $N-\theta$ relationship (dilute glycerin).

Figure 3.10. The $N-\theta$ relationship (concentrated glycerin).

200 poise; this corresponds to the stirring viscosity of SK and SF glasses. Furthermore, a concentrated millet jelly solution corresponds to the stirring viscosity of BK glasses; the screw-type stirring rod is, thus, superior in the case of such a highly viscous liquid.

Table 3.2 shows values of a and log k for crank-type and screw-type stirring rods. To eliminate striae, it is desirable to increase the stirring speed to the extent that is possible without entraining bubbles. At high speeds, the value of a determines the disappearance time θ; at low speeds, this is determined by the value of log k. The crank-type stirring rod is more efficient at low viscosities, the screw-type at high viscosities.

In regard to $N^a \cdot \theta = k$, Nagata[4] offers the following explanation: the flow velocity v of the liquid is proportional to the stirring speed n, and the circulation time τ is equal to L/v (L is the length of the circulation path). Accordingly, this can be written as $\tau = kn^{-1}$. Therefore, if the time required for the completion of mixing is considered to be the time in which mixing is completed

after m circulations, then $\theta = m\tau$. From this, $\theta = mkn^{-1}$. Rewriting this, we obtain the following relationship:

$$n\theta = mk = k' \quad .$$

Nagata et al., have written $N^a \cdot \theta = k$. However, the values of a and k vary according to the shape of the stirring rod and the viscosity of the liquid. The former is a complex flow configuration related to the rotational speed of the stirring rod, whereas the latter is a constant related to the circulation path of the liquid. In any case, switching over from clay pots to platinum pots has made it possible to select stirring rods that offer optimal stirring efficiency and has resulted in great progress in the elimination of striae.

3.1.1.2. Bubbles

Bubbles constitute one of the important factors that determine optical glass yield. It is safe to say that the elimination of bubbles is one of the greatest technical problems in melting glass, not

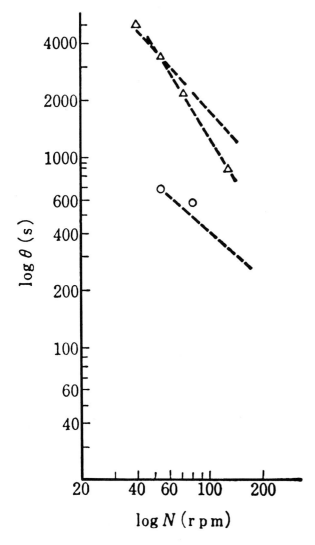

Figure 3.11. The N–θ relationship (concentrated millet-jelly solution).

Table 3.2.   Values of a and log k for crank-type and screw-type stirring rods.

		Crank type	Screw type
Dilute glycerin	a	1.8	1.2
	log k	5.5	4.4
Concentrated glycerin	a	2.3	0.7
	log k	6.8	3.8
Dilute millet jelly	a	2.1	0.9
	log k	6.3	3.9
Concentrated millet jelly	a	3.3	1.5
	log k	9.1	5.4

glass. This is interesting. The fact that bubbles in a glass are decomposition products of the raw materials suggests that the gases generated by decomposition dissolve in the melt, and that these dissolved gases form bubbles by some method.

In addition, let us consider the fining process in detail. For example, CO_2 is generated by the reaction of SiO_2 with Na_2CO_3, but because the gas is surrounded by liquid, it dissolves in the melt. Dissolution proceeds until eventually saturation is reached; if the gas diffusion in the liquid is small, a supersaturated state is formed. A gas dissolved to the point of supersaturation apparently forms bubble nuclei, and the walls of the pot and incompletely melted silica sand particles act as catalysts. It is thought that gas bubble nuclei thus generated grow into large bubbles through gas diffusion from surrounding areas; these bubbles rise through the liquid and escape from the system.

Theory of Fining.   The process of bubble formation may be thought of as a phenomenon in which a gas dissolved to the point of supersaturation in the glass pushes the liquid apart and forms a small gaseous body (i.e., a bubble) in an attempt to approach equilibrium. In other words, bubble formation may be treated as a type of phase transition phenomenon.

Formation and Growth of Bubbles in a Homogeneous System.   When a gas makes contact with a solution at the surface and reaches a state of saturation (Fig. 3.12a),

$$\mu_{II\infty} = \mu_{I0} \quad,$$

μ_{I0} is the chemical potential of one molecule of the saturated solution, and $\mu_{II\infty}$ is the chemical potential of a phase II molecule of infinitely large radius. When a state of supersaturation is reached, the supersaturated solution is in a metastable

only in optical glasses, but in all other types of glasses as well. Especially in the case of optical glasses, the presence of bubbles exceeding a maximum diameter of 50 μm cannot be tolerated. For example, when BK 7 is melted in platinum tanks, 30-μm bubbles are present at the rate of only 0.3 bubbles per kilogram.

How can bubbles in optical glasses be eliminated during melting? Let us discuss the fundamental aspects of the problem.

Causes of Bubbles.   If we analyze the bubbles in glasses, we find that they consist of gases generated by decomposition of the raw materials; they do not contain air. Of course, air is present between the particles of the raw material powders that are melted, but this air does not enter the

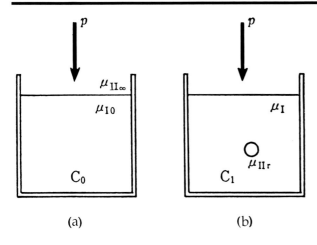

Figure 3.12.   Metastable equilibrium states between bubbles and solutions: (a) saturated solution and (b) supersaturated solution.

equilibrium with the gas bubble nuclei thus produced (Fig. 3.12b),

$$\mu_{IIr} = \mu_I \quad .$$

Here, μ_{IIr} is the chemical potential of a phase II particle that has a radius r, and μ_I is the chemical potential of the supersaturated solution. According to Volmer,[5] the relationship between the size of a particle and its chemical potential can be approximated by the following equation:

$$\mu_{IIr} - \mu_{II\infty} = \frac{2\sigma}{r} v_{II} \quad .$$

Accordingly,

$$\mu_I - \mu_{II\infty} = \frac{2\sigma}{r} v_{II} \quad . \tag{3.2}$$

In general, the equilibrium conditions that exist when a small phase II particle of radius r is in equilibrium with the surrounding phase (phase I) are given by the above equation.

If the concentrations of the saturated solution and supersaturated solution are C_0 and C, respectively, then the chemical potential of the solution is given by the following equation:

$$\mu = \mu_0 + kT \ln \gamma_C \quad .$$

Here, γ is an activity coefficient. Accordingly,

$$\mu_I - \mu_{II\infty} = \mu_I - \mu_{I0} = kT \ln \frac{\gamma_C}{\gamma_{C0}} = kT \ln \frac{C}{C_0} \tag{3.3}$$

Furthermore, if P is the external pressure, then the internal pressure of a bubble is $(P + 2\sigma/r)$. Accordingly,

$$\left(P + \frac{2\sigma}{r}\right) v_{II} = kT \quad .$$

From Eqs. (3.2) and (3.3), the radius r_k of a gas bubble nucleus, which is in equilibrium with the supersaturated solution, is given by the following equation:

$$r_k = \frac{2\sigma v_{II}}{\mu_1 - \mu_{II\infty}}$$

$$= \frac{2\sigma kT/(P + 2\sigma r)}{kT \ln C/C_0} = \frac{2\sigma}{P}\left(\frac{1 - \ln C/C_0}{\ln C/C_0}\right) \quad . \tag{3.4}$$

Equation (3.4) indicates that the size of gas bubble nuclei decreases with an increase in the degree of supersaturation C/C_0.

According to Volmer,[5] the work of nucleation A_k is given by the following equation in the case of spherical nuclei:

$$A_k = \frac{4}{3}\pi r_k^2 \sigma = \frac{16}{3}\pi\sigma^3\left(\frac{1 - \ln C/C_0}{P \ln C/C_0}\right)^2 \quad . \tag{3.5}$$

If u_1 is the activation energy of gas diffusion in the solution, then (because the gas diffusion rate is proportional to $\exp(-u_1/kT)$ the gas bubble nucleus formation rate J can be expressed as follows:

$$J = k \exp\left(\frac{-u_1}{kT}\right) \cdot \exp\left(\frac{-A_k}{kT}\right)$$

$$= K \exp\left\{-\frac{1}{kT}\left[u_1 + \frac{16\pi\sigma^3}{3}\left(\frac{1 - \ln C/C_0}{P \ln C/C_0}\right)^2\right]\right\} \quad . \tag{3.6}$$

Next, let us consider the growth of gas bubble nuclei after they have formed. In Ref. 6, Cable treats bubble growth as a diffusion phenomenon. However, Izumitani and co-workers have tried treating bubble growth as a two-dimensional phase transition (like crystal growth). The equilibrium conditions in the case of growth are given by the following equation:

$$\mu_I - \mu_{II\infty} = \sigma' \, v_{II}^{2/3} r_{k'} \quad .$$

Here, the prime corresponds to the case of a two-dimensional nucleus, and σ' is the surface energy required to form a unit length. If δ is the thickness of the two-dimensional nucleus, then we may consider that $\delta = v^{1/3}$.

The critical size $r_{k'}$ of the two-dimensional nucleus is given by the following equation:

$$r_{k'} = \frac{\sigma'}{\delta} \left(\frac{1 - 2\ln C/C_0}{P \ln C/C_0} \right) \quad .$$

The work of two-dimensional nucleation is as follows:

$$A_{k'} = \pi r_{k'} \, \sigma' = \frac{\pi \sigma'^2}{\delta} \left(\frac{1 - 2\ln C/C_0}{\ln C/C_0} \right) \quad .$$

The bubble growth rate I is given by the following equation:

$$I = k' \exp\left(\frac{-u'}{kT} \right) \cdot \exp\left(\frac{-A_{k'}}{kT} \right)$$

$$= K' \exp\left[-\frac{1}{kT} \left(u' + \frac{\pi \sigma'^2}{\delta} \frac{1 - 2\ln C/C_0}{P \ln C/C_0} \right) \right] .$$

$$(3.7)$$

If the gas diffusion rate is assumed to be inversely proportional to the viscosity, then

$$I = K' \left(\frac{1}{\eta} \right) \exp\left(\frac{-A_{k'}}{kT} \right) \quad .$$

Furthermore, with A and B as constants at a fixed temperature,

$$\log I = A - B \frac{1 - 2\ln C/C_0}{P \ln C/C_0} \quad .$$

From Eqs. (3.6) and (3.7), we may conclude that bubble nucleation and bubble growth rates increase with a decrease in surface energy, an increase in the degree of supersaturation, or an increase in the gas diffusion rate.

Formation of Gas Bubble Nuclei in a Heterogeneous System. How does the rate of gas bubble nucleus formation in a heterogeneous systems (e.g., a system in which a solid surface is in contact with a supersaturated solution) compare with that found in a homogeneous system? This rate is affected by the "wetting" of the solid surface. Now, assuming that A_{k*} is the work of nucleation

$$A_{k*} = \Delta\mu_v \cdot V_2 + \sigma_{12} \cdot S_{12} + \left(\sigma_{23} - \sigma_{13} \right) S_{23} \quad .$$

$$(3.8)$$

Here, $\Delta\mu_v = (\mu_{II\infty} - \mu_I/v_{II})$; this is the difference in chemical potential (per unit volume) between phases I and II. The term V_2 is the volume of the gas bubble nucleus, σ is the surface energy, and S is the surface area. As for the subscripts, 1 indicates the solution, 2 indicates the gas bubble nucleus, and 3 indicates the solid; Fig. 3.13 shows the relationship of the three.

Now, because $\sigma_{13} = \sigma_{23} + \sigma_{12} \cos (\pi - \theta)$,

$$\sigma_{23} - \sigma_{13} = \sigma_{12} \cos \theta \quad ,$$

$$S_{12} = 2 \pi r^2 \left[1 - \cos (\pi - \theta) \right]$$

$$S_{23} = \pi \left[r \sin (\pi - \theta)^2 \right]^2$$

$$V_2 = \frac{1}{3} \pi r^3 \left[2 - 3 \cos (\pi - \theta) + \cos^3 (\pi - \theta) \right] .$$

Furthermore, the size of the gas bubble nucleus is given by

$$r^* = - 2 \sigma_{12}/\Delta\mu_v \quad .$$

Accordingly, if we substitute these values into Eq. (3.8), and make the necessary adjustments, we obtain the following equation:

$$A_{k*} = \frac{1}{3} \pi r_k^2 \, \sigma_{12} (2 - \cos \theta)(1 + \cos \theta)^2 \quad .$$

Now, if we write $A_{k*} = A_k \cdot f(\theta)$, then

$$f(\theta) = \frac{1}{4} (2 - \cos \theta)(1 + \cos \theta)^2 \quad .$$

The term $f(\theta)$ is a function of the angle of wetting θ; when $\theta = 0$, $f(\theta) = 1$, and when $\theta = \pi$, $f(\theta) = 0$. As shown by Fig. 3.14, $f(\theta)$ decreases as the angle of wetting increases. In other words, as the solid surface becomes harder to wet, A_{k*} decreases, and gas bubble nucleus formation tends to occur more easily.

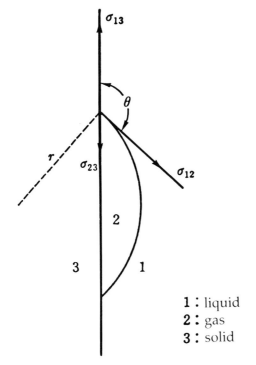

1 : liquid
2 : gas
3 : solid

Figure 3.13. Nucleation in a heterogeneous system.

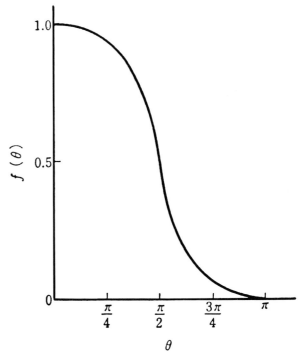

Figure 3.14. Relationship between angle of wetting and $f(\theta)$.

The Glass Fining Process. The above mentioned theoretical equations concerning bubble formation and growth have been confirmed by dissolving carbon dioxide in water and glycerin to the point of supersaturation and by measuring the rate of formation of the bubbles thus generated.[7] However, the question of whether or not bubble generation occurring in the glass fining process can be treated as a supersaturation phenomenon was investigated as described below.

Bubble Growth. The number of gas bubbles in a BK 7 melt and temporal changes in the mean diameter of these bubbles were measured to ascertain whether or not a bubble growth process occurred in the glass fining process. The BK 7 specimens were maintained at 1450°C in platinum pots for various periods of time. Afterward, the specimens were cast in cast-iron molds. The number and sizes of bubbles in the samples thus obtained were measured, and the results of this measurement are shown in Fig. 3.15. The figure shows results for a carbonate batch, and a batch in which a portion of the carbonate was replaced by nitrate. In both cases, the number of bubbles decreased exponentially with time. The mean diameter of the bubbles, on the other hand, increased immediately after melting, reached a maximum value and

then decreased (as shown in Fig. 3.15). This indicates that bubbles in glass develop as follows: first, large numbers of relatively small bubbles appear. These bubbles gradually grow larger; as large bubbles, they rise through the melt and escape from the system. Figure 3.15 shows that a bubble growth process occurs in the glass fining process and indicates that the generation and growth of bubbles in glass may be treated as a type of phase transition phenomena.

Effect of Degree of Gas Supersaturation. The effect of the degree of gas supersaturation was investigated for BK 7 and lanthanum borate glasses from the standpoint of the relationship between the amount of dissolved gas (using batches and cullets) and the number of bubbles. In all cases, the batches (which contained large amounts of dissolved gas) showed higher bubble removal rates. Table 3.3 shows one example of this. In the case of the lanthanum borate glass, bubbles were completely absent after 80 minutes at 1200°C in batch melting, whereas bubbles were still present even after 160 minutes in cullet melting. The amounts of gas extracted from the glass after 40 minutes of melting were 0.96 and 0.12 ml/g, respectively; thus, the batch-melted glass had dissolved a greater amount of gas. Because the amount of dissolved gas was small in the case of cullet melting, the bubbles could not grow, and small bubbles probably remained indefinitely.

Effect of Gas Diffusion Rate. Stirring increases the cross-sectional area of bubble diffusion, and also increases the concentration gradient. Accordingly, stirring would be expected to cause bubble growth so that the bubble removal rate is accelerated. Table 3.4 shows an example of this. This table shows results for BK 7; the glass was melted at 1450°C in a two-liter platinum pot, and stirring was performed for 30 minutes beginning 110 minutes after batch feeding. Stirring caused an increase in bubble diameter, and a sharp decrease in the number of bubbles.

The facts in Table 3.4 confirm that a bubble growth process occurs in the bubble fining process, and that the bubble growth rate increases (so that bubble removal is accelerated) with an increase in the degree of supersaturation or an increase in the rate of diffusion.

Bubble Removal Techniques for Optical Glass. Among the approximately 130 types of optical glasses, those that show large numbers of bubbles include high-viscosity glasses such as KF glasses and extremely low-viscosity glasses such as La and SSK glasses. In high-viscosity glasses, relatively small numbers of large bubbles tend to remain in the glass; minute bubbles (around 10 μm) are absent. In extremely low-viscosity glasses, on the other hand, extremely minute bubbles (10 μm or smaller) are often generated. Furthermore, the presence of residual bubbles in the glass when the glass is removed and examined after stirring (even when the bubbles have disappeared at the fining

Table 3.3. Difference in bubble removal rate between batches and cullets (number of bubbles).

Time (min)	40	80	120	160
Batch (100 ml)	0.8	0	0	0
Cullet (100 ml)	3.2	1.0	4.4	1.8

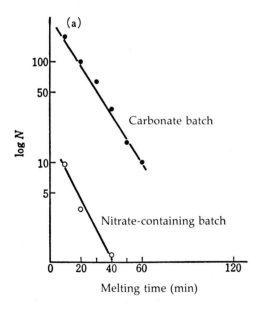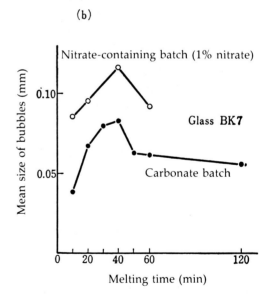

Figure 3.15. Relationship of melting time to (a) number of bubbles and (b) mean size of bubbles.

Table 3.4. Effect of stirring on fining (number of bubbles of various sizes).

	>0.25 (mm)	0.25 to 0.15 (mm)	0.15 to 0.06 (mm)	0.06 to 0.03 (mm)	<0.03 (mm)	Total number	Mean size (mm)
Before stirring	0.4	0.76	22.6	60.8	44.1	135.5	0.060
After stirring	1.0	0	0.2	1.0	1.0	3.2	0.167

temperature) is something that is often experienced. Let us discuss bubbles in SK 1 as an example of the latter phenomenon, and bubbles in SSK 2 as an example of minute bubbles.

Bubbles in SK 1. The composition (by weight) of SK 1 is 40.1% SiO_2, 5.7% B_2O_3, 2.5% Al_2O_3, 8.5% ZnO, 42.2% BaO, 0.5% PbO, and 0.5% As_2O_3. This glass was allowed to stand at the fining temperature until it was confirmed that almost no bubbles were visible. The glass was then cooled and subjected to a final stirring; when the glass was removed from the furnace afterward, large numbers of bubbles were present. No bubbles were observed when this glass was cooled without final stirring. More bubbles were generated when a carbonate was used as a raw material; fewer bubbles were generated when a nitrate was used. Figure 3.16 shows the solubility curves of O_2 and CO_2. Table 3.5 lists the amounts of dissolved gas.

The solubilities of O_2 and CO_2 decrease with a rise in temperature, and CO_2 has a higher solubility than O_2. Furthermore, the amount of dissolved gas and degree of supersaturation are also greater when a carbonate is used than when a nitrate is used. This may be interpreted as follows: when a carbonate is used, the amount of dissolved gas is large and the degree of supersaturation is high; as a result, bubbles are generated by stirring.

Bubbles in SSK 2. The composition (by weight) of SSK 2 is 37.3% SiO_2, 6.3% B_2O_3, 2.5% Al_2O_3, 8.0% ZnO, 39.7% BaO, 4.0% PbO, and 1.0% As_2O_3. The glass SSK 2 has a lower viscosity than SK 1. Samples of SSK 2 were melted at 1400°C in two-liter platinum pots; when a nitrate was used as a raw material, minute bubbles (around 10 μm) were present whether stirring was performed or not. However, by switching the raw material from nitrate to carbonate, it was possible to eliminate small bubbles (10 μm or smaller). Because SSK 2 has a low viscosity and a high O_2 diffusion rate, bubbles are generated even at a low degree of supersaturation. As a result of this, the degree of supersaturation drops even further. It appears, therefore, that the bubbles cannot grow because of the low degree of supersaturation, and thus remain as minute bubbles. According to Cable,[6] the diffusion rates of O_2 and CO_2 are 3×10^{-6} cm²/s and 5×10^{-7} cm²/s, respectively. The diffusion rate of O_2 is faster than that of CO_2; accordingly, when a nitrate is used as a raw material, bubbles are formed rapidly, and the amount of dissolved gas decreases abruptly. On the other hand, when a carbonate is used as a raw material,

the diffusion rate of CO_2 is slow so that bubble formation is retarded. The CO_2 remains dissolved in the glass so that the degree of supersaturation is maintained. Accordingly, the bubbles grow and are eliminated from the system.

Relationship between Degree of Supersaturation and Bubbles. In the previous section, equations for the formation rate and growth rate of bubbles were derived. Now, assuming that $P = 10^6$ dyne/cm², $\sigma = 300$ dyne/cm, and $\delta = 10^{-8}$ cm in a case where $C/C_0 = 1.2$, the work required for bubble growth (for one molecule) is

$$A_{k'} = \pi r_{k'} \sigma'$$

$$= \frac{\pi \sigma'^2}{P\delta} \ln \frac{C}{C_0} \simeq 2.7 \times 10^{-11} \text{ erg} \quad ,$$

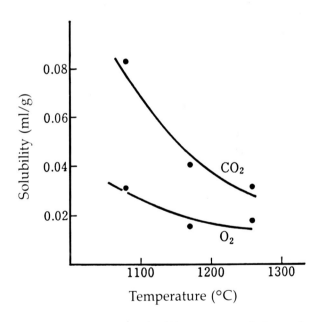

Figure 3.16. The solubility curves of O_2 and CO_2 (SK 1).

Table 3.5. Amounts of dissolved CO_2 and O_2 in SK 1.

	Amount of dissolved gas (ml/g)	C/C_0
Carbonate	0.075	2.3
Nitrate	0.025	1.6

and the work required for bubble formation

$$A_k = \frac{4}{3} \pi r_k^2 \sigma$$

$$= \frac{16 \pi \sigma^3}{3P^2} \left(\ln \frac{C}{C_0} \right)^2 \sim 2.4 \times 10^{-4} \text{ erg} \quad .$$

Thus the activation energy of nucleation is far greater than the activation energy of growth. Consequently, a greater degree of supersaturation is required for bubble generation than for bubble growth. Figure 3.17 is a model that shows the activation energies of formation and growth, in addition to the rates of formation and growth, as functions of $\ln C/C_0$. In the case of a low degree of supersaturation (and in the case of a high viscosity), a region exists where nucleation does not occur but growth can occur (region b). As the degree of supersaturation increases, both the number and size of bubbles increase (region c). It appears that when a nitrate is used in SSK 2, bubbles are easily formed because of the high O_2 diffusion rate; as a result, the degree of supersaturation drops abruptly so that bubble growth cannot occur (region a). When a carbonate is used, on the other hand, there is little bubble formation because of the low diffusion rate of CO_2; accordingly, the degree of supersaturation is maintained at a relatively high level so that the bubbles grow and are eliminated (region b). When a carbonate is used in SK 1, the degree of supersaturation is high (region c) so that bubbles are formed by stirring. In high-viscosity glasses such as BK, KF, etc., there is also little bubble formation; in such glasses, however,

it appears that the degree of supersaturation is maintained at a high level so that it is more advantageous to use a nitrate. The fact that the bubbles in SSK glasses are small, whereas the bubbles in BK 7 are large, is also explained by Fig. 3.17. In the case of BK 7, etc., the viscosity is high; as a result, bubbles are formed wherever the degree of supersaturation is relatively high, but the number of bubbles is small, and only the growth of these bubbles proceeds so that no small bubbles are present. In SSK glasses, on the other hand, the viscosity is low; as a result, bubbles are formed even at a low degree of supersaturation. However, because the degree of supersaturation is low, these bubbles do not grow, and many small bubbles are generated. Thus, to eliminate bubbles in optical glasses, it is extremely important to know the types of gases present in the glass, and the degrees of supersaturation of these gases.

Bubble Removal Techniques. There are two methods for removing bubbles. One is to maintain the melt at a relatively high temperature so that bubbles are caused to grow and float upward. The other is to maintain the melt at a relatively low temperature so that small bubbles are eliminated by dissolution. The latter method is commonly used to eliminate bubbles in flint glasses. The basic factors that must be considered in these methods are the rates of diffusion and the solubility curves of the dissolved gases. The diffusion rates and solubilities of gases such as CO_2, O_2, H_2O, N_2, etc., at various temperatures and the relationships of these factors to glass compositions must be determined. The gases H_2O, CO_2, and O_2 undergo chemical dissolution; the

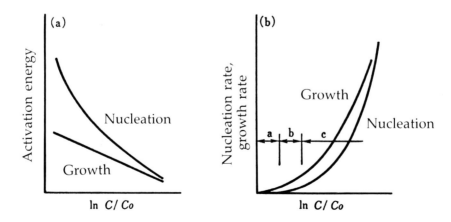

Figure 3.17.   Relationship of degree of gas supersaturation to (a) activation energies of bubble nucleation and growth and (b) rates of bubble nucleation and growth.

solubility decreases with a rise in temperature. Furthermore, N_2 gas generally undergoes physical dissolution so that the solubility increases with a rise in temperature. In the case of the former gases, therefore, bubbles can be squeezed out; whereas in the latter case, lowering the temperature causes bubbles to evolve.

The fining of optical glasses is performed as follows: Figure 3.18 shows that the degree of supersaturation of the dissolved gas is high immediately after melting; the melt is in state (1) at a temperature of T_1. In the glass melt in state (1), stirring causes the formation and growth of bubbles; as a result, bubbles are removed from the melt and the melt is degassed at the same time, and shifts to state (2). When the glass is cooled so that it reaches temperature T_2, the melt enters an unsaturated state (because of the solubility curve) so that any bubbles that were previously present are absorbed by dissolution. At this temperature, of course, no bubbles will be produced even if the melt is stirred. Thus, a certain amount of gas is required to cause bubble growth. If the amount of gas is too small, minute bubbles will form, and degassing will become difficult in high-viscosity glasses. If degassing is insufficient, the final stirring will create bubbles. Furthermore, the vessel walls and the fining agent play important roles in glass bubble formation. Platinum pots are more

wettable than refractory pots so that bubbles do not tend to be generated. When a fining agent such as As_2O_3, Sb_2O_3, etc., is used together with a nitrate, O_2 bubbles are generated; these bubbles absorb the dissolved gas in the glass melt, grow to a large diameter, float upward, and leave the system.

The elimination of bubbles in the approximately 130 types of optical glass is not a simple task; it has to be done on a case-by-case basis. The principle, however, is simple (as shown in Fig. 3.18); still, basic data are scarce. The solubilities (degrees of supersaturation) of various gases, relationships between glass compositions and gas solubilities, and diffusion rates of various gases in glasses (and especially the two former items) need to be known for the fining of glasses. The accumulation of data is essential. Figure 3.19 shows amounts of dissolved gas for several glasses. The amount of dissolved gas increases with an increase in viscosity.

3.1.2. Pressing

The manufacture of optical glasses includes two processes: melting and forming. Melting has progressed to such an extent that once a batch has been inserted, glass that is free of striae or bubbles

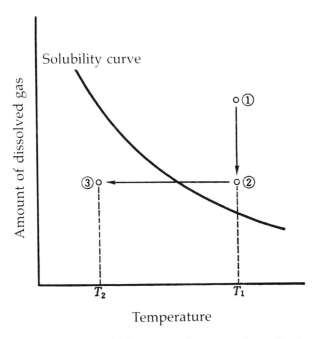

Figure 3.18. Bubble removal process in optical glasses.

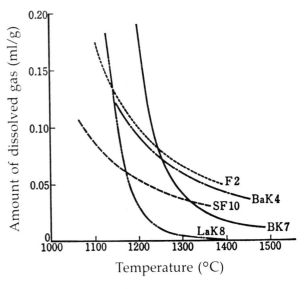

Figure 3.19. Amounts of dissolved gas in optical glasses.

is fed out continuously. Well then, how about forming?

In the field of forming as well, a direct pressing method has been developed in which glass exiting after bubble removal and homogenization is immediately cut and pressed. Until this method was developed, the usual practice was as follows.

Only homogeneous portions of glass melted in pots were selected, or glass formed in the shape of a continuous E-bar (extruded bar) was cut. After the weight of the glass was adjusted by barrel grinding, the glass was softened by being passed through an electric furnace, and was then hand pressed. This process is shown in Table 3.6. In

Table 3.6.   Lens blank manufacturing processes.

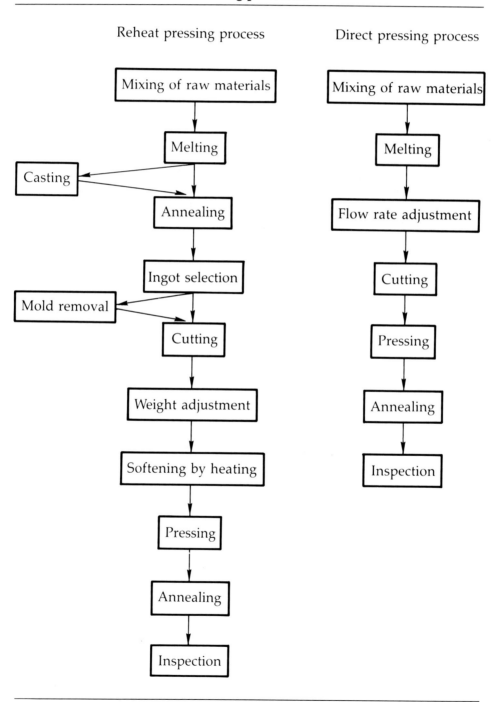

direct pressing, the molten glass is not cooled; glass flowing out of a platinum tube is immediately cut and pressed. Consequently, the process is greatly simplified and costs are reduced. Furthermore, by linking this process with direct annealing, the time required for manufacture can be greatly reduced. Moreover, because the weight can be precisely adjusted, a great improvement is seen in the precision of lens thicknesses, diameters, etc.

3.1.2.1. Reheat Pressing

In reheat pressing, the selected optical glass ingots are ordinarily cut, adjusted for weight by means of a balance, and sent to the pressing process. A problem that arises here is nonuniformity of the cut weight. This causes fluctuations in thickness. Furthermore, scratches that occur at the time of cutting cause bubbles (mist*) in the pressed products. In addition, nonuniformity of the shape of the cut pieces leads to tucking.[†] The raw glass ingots used for pressing are supplied in the form of irregular lumps, strips (sheets of indefinite length that are rectangular in cross section), or gobs, etc. Because strips have a fixed shape, and gobs have a fixed shape and weight, these forms are convenient for re-pressing. In the case of gobs, there is no need for weight adjustment, and the drawbacks of misting and tucking do not occur. Accordingly, short-footed** glasses that cannot be direct-pressed are most advantageously pressed in the form of gobs.

In the case of reheat pressing, in addition to the aforementioned weight adjustment and occurrence of scratching, there is also the problem of selecting an appropriate mold-releasing agent.

A state in which soft parts of a glass overlie hard parts when the glass is softened or formed is called tucking. Misting refers to the generation of small bubbles in boundary areas when scratches in the cut piece are filled during the softening process. Scratches may be classified as follows: scratches that were present from the beginning in the glass ingot, scratches that were generated in the cutting process, and scratches that were generated during handling of the cut piece. An attempt is made to remove scratches by barrel working, etc., during the weight adjustment process; how-

ever, although scratches in corner areas are easy to eliminate, scratches in flat areas may be difficult to remove. For this reason, it is sometimes necessary to remove as much as 0.5 mm in the rough grinding process. The need to remove as much as 0.5 mm in rough grinding is attributable to the mold releasing agent and shrinkage, as well as to misting. In the case of gobs, the problem of misting is resolved.

In the case of reheat pressing, the glass piece is placed on a refractory pan that is sprinkled with a mold releasing agent (alumina, kaolin, BN, etc.) or coated with a paste of such an agent; and the glass is then softened. In this way, fusion of the glass to the refractory is prevented, and transfer of the glass piece to the metal pressing dies is facilitated. However, the use of such a mold releasing agent causes the glass surface to be covered by an opaque powder so that observation of the interior of the pressed item is difficult. Furthermore, this powder must be removed by the lens maker in a rough grinding process. The problem of glass and mold releasing agent is a problem of the wetting of the glass melt and the mold releasing agent powder. In recent years, it has become possible to obtain a semitransparent surface using boron nitride.

3.1.2.2. Direct Pressing

Direct-pressed products have the final form of the product in question; there is no weight adjustment process and no misting. However, such products do suffer from the defects of gaps and share marks. Share marks consist of groups of small bubbles that can be observed with a microscope. These bubbles extend into the interior of the glass to a depth of approximately 0.3 mm from the surface. Figure 3.20 shows this phenomenon. The minimization or elimination of this infiltration is an important technical problem. In the case of direct-pressed products, the depth of the layer removed by rough grinding is determined by the depth of these share marks.

Defects that are shared by both direct pressing and reheat pressing are internal product defects such as waver* and gaps[†], etc., and defects in the precision of external dimensions (thickness defects, external diameter defects, defects in radius of curvature, deformation, etc.).

---

* Small bubbles generated by reheating at the contact of two cracked surfaces.

[†] Interface formed when the original surface is folded into the interior of the pressed piece.

** Steeper viscosity curve.

---

* Straight crack thermally generated from the contact of the mold with cooled glass.

[†] Crooked surface cracks thermally caused by contact of glass with a low-temperature mold.

Heat Transfer and Glass Solidification during Pressing—Waver and Deformation. What kind of phenomenon is glass pressing? The pressing of glass consists of two functions: the forming of a shape and the removal of heat from the high-temperature glass. Accordingly, the temperature of the glass at the time of forming must be high enough to cause viscous flow. Direct pressing is performed at a temperature corresponding to about 10^3 poise, whereas reheat pressing is performed at a temperature corresponding to about 10^5 poise. The pressing pressure is 1 to 20 kg/cm². The next thing that must be considered is the heat transfer from the glass to the metal mold during pressing. As shown by Table 3.7, the thermal conductivities of glasses are extremely low. Accordingly, the rate at which heat is removed is governed by the thermal conductivity of the glass so that its temperature does not tend to drop rapidly.

Many studies[8] have been conducted concerning the temperature distribution and amount of heat transfer, etc., inside the glass, at the interface, and inside the metal mold during forming. The problem here is that the heat transfer coefficient of the interface between the glass and the metal mold is not fixed. The value of this coefficient depends on the tightness of the interface between the pressed glass and the metal mold, and also varies according to pressure. Figure 3.21 shows the temperature distribution inside the glass during glass forming. The surface temperature is low, whereas internal temperatures are high. Therefore, when the pressing time is several seconds and the glass is removed from the mold with the temperature at the center of the glass still at a high level and this glass is then cooled, it is difficult to obtain fixed dimensions and the pressed shape is not uniform. This case also leads to shrinkage.

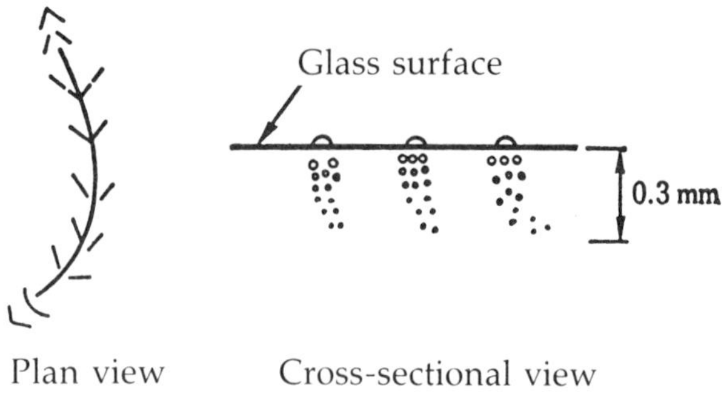

Plan view   Cross-sectional view

Glass surface

0.3 mm

Figure 3.20.   Model diagram of share marks.

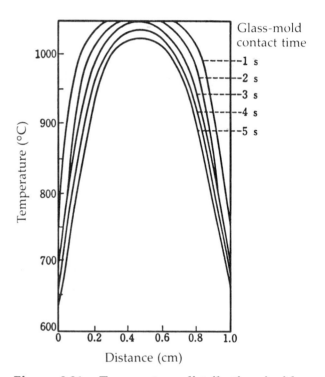

Figure 3.21.   Temperature distribution inside glass during pressing.

Table 3.7.   Physical properties of metal-mold materials and glasses (20°C).

Substance	Density (g/ml)	Specific heat (cal/g·°C)	Thermal conductivity (kcal/m·h·°C)
Steel	8.9	0.100	320
Cast iron	7.27	0.100	41
Chromium steel	7.79	0.110	27
Lime glass	2.21	0.170	1.16
Soda glass	2.59	0.180	0.64
Flint glass	3.45	0.120	0.52

When the glass and the metal surface are in perfect contact, the temperature at the glass interface is given by the following equation[9]:

$$\frac{\theta_G - \theta_S}{\theta_S - \theta_M} = \frac{\sqrt{\lambda_M C_M \rho_M}}{\sqrt{\lambda_G C_G \rho_G}} \quad .$$

Here, θ_G is the initial temperature of the glass (gob), θ_M is the initial temperature of the metal mold, θ_S is the glass surface temperature, λ is the thermal conductivity, C is the specific heat, and ρ is the specific gravity.

Immediately after pressing, it is thought that contact is close to perfect: as time passes, however, it is thought that contact becomes poorer. Consequently, calculating the interface temperature accurately is difficult. In general,* as shown in Fig. 3.21, the interior of the glass is in a liquid state, while the surface layer is abruptly cooled and solidified during pressing. In the solidified layer, a temperature difference exists between the surface and the interior part; accordingly, the surface is subjected to tension. Therefore, if the metal-mold temperature is too low, so that the glass- and metal-mold interface temperature is low and the magnitude of the aforementioned tension is greater than the strength of the glass, the glass will break. This is called waver and gaps. Conversely, if the metal-mold temperature is too high, the glass and the metal mold will fuse together. Or, deformation will tend to occur after pressing. Thus, an optimal value exists for the metal-mold temperature. Because the temperature-viscosity curve varies according to the type of glass involved, the solidification temperature will vary even at the same fixed cooling rate. Accordingly, the metal-mold temperature must be selected according to the type of glass involved. In general, the more short-footed the glass, the greater the need to increase the metal-mold temperature to prevent breaking during pressing. Thus, waver and deformation are the most prevalent defects in the glass pressing process.

Dimensional Precision. In pressed products, the dimensions where dimensional precision is a problem are thickness, external diameter, and radius of curvature. Of these, thickness is determined by the weight of the glass that enters the

metal mold, whereas the precision of the external diameter and radius of curvature is determined by the molding conditions.

Thickness. The external diameter of a pressed procuct is determined by the internal diameter of the mold. Because the thickness of the pressed item is proportional to the weight of the glass in the mold, the dimensional precision of the thickness is determined by the precision with which the glass weight is cut. Reheat pressing depends on a balance and manual handling; therefore, the weight fluctuation is controlled to within 10%.

In the case of direct pressing, molten glass flowing out of an outflow tube is automatically cut at fixed time intervals. Accordingly, the minimization of weight fluctuation depends on the stability of the flow rate. The flow rate of the glass flowing out of the outflow tube is given by the Hagen-Poiseuille equation:

$$Q = \frac{\pi r^4 \rho g (h + l)}{8 \mu l} \quad .$$

Here, h is the fluid head, l is the length of the outflow tube, r is the radius of the outflow tube, and μ is the viscosity. The terms r and l are fixed, being determined by the apparatus. If the glass temperature and fluid head are maintained at constant values, the flow rate will stabilize so that the weight fluctuation can be controlled to within 3%.

External Diameter. The external diameter of a pressed product is calculated with respect to the dimensions of the metal mold by the following equation:

$$d = d_0 \left[1 + \left(T_M - T_1 \right) \alpha_0 \right]$$
$$\left[1 - \left(T_G - T_g \right) \alpha_1 \right]\left[1 - \left(T_g - T_1 \right) \alpha_2 \right] \quad .$$

Here, d is the dimension of the pressed product, d_0 is the corresponding dimension of the mold, T_1 is room temperature, T_M is the mold temperature, T_G is the temperature of the glass on completion of pressing, T_g is the glass transition temperature, α_0 is the expansion coefficient of the mold material, α_1 is the expansion coefficient of the glass above T_g, and α_2 is the expansion coefficient of the glass below T_g.

There is a temperature distribution in T_G; even if the metal-mold temperature, gob temperature, and pressing time are fixed, it is difficult to determine T_G with accuracy, and therefore difficult to obtain an accurate measured value of α_1.

---

* The gob temperature and metal-mold temperature are adjusted so that the interface temperature will be in the vicinity of the yield point of the glass.

For this reason, the relationship between the dimensions of the pressed product and the dimensions of the metal mold must depend on experimental values. Figure 3.22 shows the relationship between the amount of shrinkage and the dimensions of the pressed product. It is difficult to fix accurate dimensions for the pressed product in mold design. However, if the metal-mold temperature, gob temperature, pressing time, and pressing pressure are fixed, the temperature of the glass on completion of pressing will be fixed so that fluctuations in external diameter can be controlled. The gob temperature affects T_G directly; the temperature of the mold has an effect on T_G and on the dimensions of the mold during forming. The temperature T_G drops as the pressing time is increased; as a result, the external diameter increases. The variation of external diameter with pressing time is shown in Fig. 3.23. Table 3.8 shows working tolerances for pressed products.

Radius of Curvature. The curvature radius of a pressed product is determined by the curvature radius of the metal mold. However, there is a temperature distribution during pressing so that a difference exists between the surface layer and the interior in terms of the amount of shrinkage that occurs in the cooling process after forming. As a result, the surface is pulled inward so that shrinkage occurs. Shrinkage tends to occur more readily in direct pressing, where the gob temperature is higher, than in reheat pressing.

3.1.3. Annealing

Lens blanks that have been homogeneously melted and pressed are annealed. The annealing process is an important process that might be called the final finishing in the manufacture of optical glasses. This process serves the following four purposes:
1. Eliminates residual strain in the glass.

2. Imparts the same index of refraction to the entire glass specimen (homogenization).

3. Eliminates variation in index of refraction between specimens [this is accomplished by treating specimens with different thermal histories (because of pressing, etc.) at a fixed temperature so that the specimens are stabilized].

4. Causes the index of refraction of the product to agree with the stipulated optical constants (this can be achieved to some extent by adjusting the soaking temperature, cooling rate, etc.).

Two methods exist for the annealing of optical glasses (Fig. 3.24). These are the constant-rate cooling method and the constant-temperature soaking method. The former method aims at stabilization and uniformity of the index of refraction by maintaining the glass at a relatively high fixed temperature (e.g., T_g) and then cooling the glass at a constant rate so that the glass structure is frozen at a fixed structural temperature. In the latter method, thermal residual stress is eliminated and the structure is stabilized by maintaining the glass at a constant temperature for a long period of time; afterward, the glass is gradually cooled so that no stress enters the glass. Today, the former method is employed almost exclusively because of the advantages it offers in terms of homogeneity and total treatment time. As shown in Fig. 3.25, annealing consists of a heating process, a constant-temperature soaking process, a constant-rate cooling process, and a rapid cooling process. Of these, the most important processes are the constant-temperature soaking process and the constant-rate cooling process. We will now discuss in some detail how the index of refraction varies and how residual stress is relaxed in these processes.

It is known that when a glass is maintained at a temperature in the transition region, the structural temperature of the glass changes and approaches a fixed value. According to Tool,[10]

Table 3.8. Working tolerances of pressed products.

External diameter (mm)	Thickness tolerance (mm)		External diameter tolerance (mm)	
	Re-pressing	Direct pressing	Re-pressing	Direct pressing
<10	±0.5	0.5	±0.1	0.2
10 to 30	±0.4	0.4	±0.1	0.2
30 to 50	±0.4	0.4	±0.2	0.3
50 to 70	±0.4	0.4	±0.3	0.3
70 to 100	±0.5	0.5	±0.4	0.4
>100	±0.6	0.6	±0.5	0.5

$$\frac{d\tau}{dt} = K(T - \tau)\exp\left(\frac{T}{g}\right)\cdot\exp\left(\frac{\tau}{h}\right) \quad,$$

where $T - \tau$ is a force term, and $K\exp(T/g)\cdot\exp(\tau/h)$ is a flow term. Here, as τ approaches T, the above equation can be written as follows:

$$\frac{d\tau}{dt} = K_T(T - \tau) \quad.$$

For example, Fig. 3.26 shows the change in viscosity over time when glass specimens are maintained at constant temperature. This change over time can be described as a change in structural temperature.[11] Viscous flow is not simply a function of temperature alone; it changes over time as

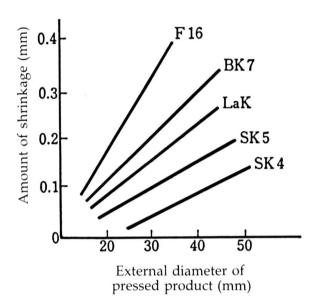

Figure 3.22. Relationship between dimensions of the pressed product and amount of shrinkage.

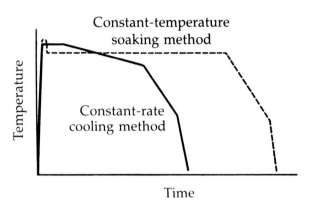

Figure 3.24. Annealing methods.

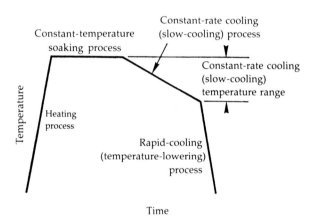

Figure 3.25. Various processes in the annealing schedule.

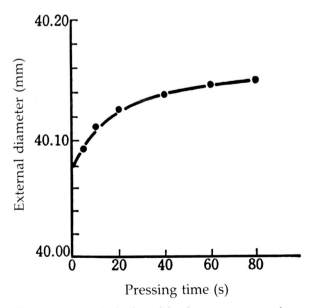

Figure 3.23. Relationship between pressing time and external diameter.

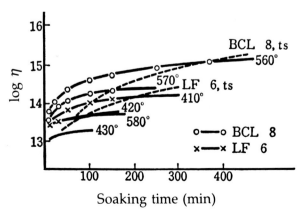

Figure 3.26. Changes in viscosity when samples are soaked at fixed temperatures.

a change in structural temperature. This is the reason that the residual stress relaxation process and changes in index of refraction over time show complex behavior. Moreover, the relaxation process in the transition region cannot be expressed by a viscosity term alone; processes with a shorter relaxation time (e.g., delayed elasticity) also make a contribution. Accordingly, it is not easy to describe changes in the properties of a glass in the transition region. It is necessary to consider two or three relaxation mechanisms.

The glass samples used here were LF 6 (as an example of conventional silicate glasses) and BCL 8 (B_2O_3–CdO–La_2O_3 glass, used as an example of new lanthanum glasses). The thermal characteristics of these glasses are shown in Table 3.9.

3.1.3.1.   Changes in Index of Refraction Caused by Heat Treatment

Changes in the Index of Refraction Resulting from Soaking at a Fixed Temperature. In regard to changes in the index of refraction resulting from soaking at a fixed temperature, the following three equations are known:

$$n = n_e + (n_0 - n_e) \cdot \exp(-AT) \quad , \qquad (3.9)$$

$$\frac{1}{n_e - n} - \frac{1}{n_e - n_0} = At \quad , \qquad (3.10)$$

$$\frac{dn}{dt} = Q(n_e - n)\exp\left(\frac{n}{P}\right) \cdot \exp\left(\frac{T}{g}\right) \quad . \qquad (3.11)$$

Equation (3.9) is called Winter's equation (Ref. 12), and corresponds to Maxwell's equation for stress relaxation. Equation (3.10) is McMaster's equation (Ref. 13), and corresponds to the stress relaxation equation of Adams and Williamson. Equation (3.11) is Collyer's equation (Ref. 14), and corresponds to Tool's structural temperature equation. Furthermore, the equation of Adams and Williamson can be derived using Maxwell's equation when the viscosity varies linearly with time at a fixed temperature (so that $\eta = \eta_0 + at$).

Figures 3.27 and 3.28 show the equilibrium indices of refraction for the respective glasses in the vicinity of T_g, and the time (t_s) required to reach these equilibrium indices of refraction. The results obtained when changes in the index of refraction of BCL 8 over time were treated using Winter's equation are shown in Fig. 3.29. When the initial processes of abrupt change are treated using Collyer's equation, the results are shown in Fig. 3.30. When the relationships beween $dn/dt/n_e - n$ and $1/T$ for lines of fixed index of refraction are determined from this figure, the results are as shown in Fig. 3.31, and an activation energy of 114 kcal/mol can be obtained for the initial stage. Furthermore, if the second-stage activation energy is determined acording to Winter's equation, a value of 223 kcal/mol is obtained. Similarly, an initial-stage value of 39.8 kcal/mol and a second-stage value of 67.2 kcal/mol were obtained for LF 6. These results indicate the existence of two different relaxation mechanisms.

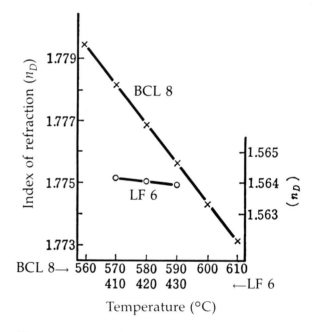

Figure 3.27.   Relationship between equilibrium index of refraction and temperature.

Table 3.9.   Characteristic values of LF 6 and BCL 8.

	Softening point ($10^{7.6}$ poise)	Yield point (10^{10} poise)	Transition point ($10^{13.5}$ poise)	Strain point ($10^{14.5}$ poise)
BCL 8	668°C	610°C	578°C	566°C
LF 6	630°C	479°C	420°C	390°C

Changes in Index of Refraction Caused by Constant-Rate Cooling.

Cases Where the Glass is Cooled from a Fixed Soaking Temperature. Figure 3.32 shows the following: the changes in the index of refraction that occurred when BCL 8 was cooled at rates of 4.67, 15, and 48°C/hr after soaking at 610°C; the relationship between equilibrium index of refraction and temperature; and the changes in the

index of refraction that occurred when the glass was cooled at a rate of 15°C/hr after soaking at 586 and 580°C.

● Relationship between cooling rate and T_R: when BCL 8 is cooled at a constant rate, the index

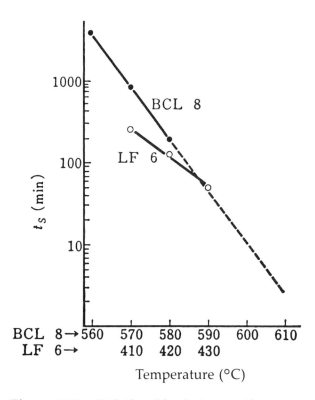

Figure 3.28. Relationship between time required to reach equilibrium and temperature.

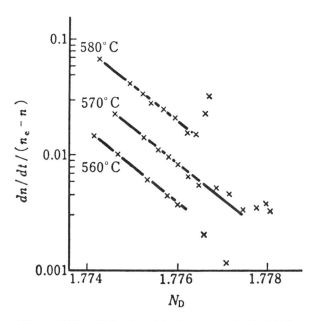

Figure 3.30. Relationship between ln($dn/dt/n_e - n$) and index of refraction.

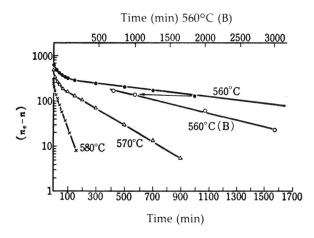

Figure 3.29. Relationship between ln($n_e - n$) and t (BCL 8).

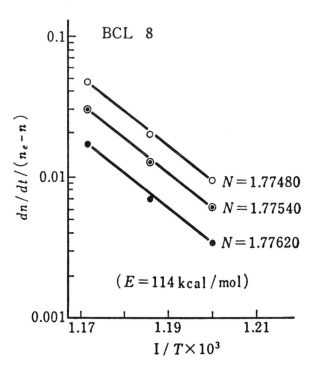

Figure 3.31. Relationship between ln($dn/dt/n_e - n$) and $1/T$.

of refraction initially varies linearly; when a certain point is reached, however, the variation begins to diverge from this straight line. If this temperature is called T_R, the T_R varies according to the cooling rate, and (as shown in Fig. 3.33) this relationship can be expressed by Ritland's equation (Ref. 15): $T_R = T_1 + K \ln R$. In terms of index of refraction, the relationship is indicated by the chain line in Fig. 3.33. It is seen that T_R shows a divergence from τ_R, which is determined from the equilibrium index of refraction.

• Concerning the index variation that occurs below T_R: in regard to the index variation that occurs below T_R, the relationship between $n_R - n_T$ and cooling rate (where n_T is the final index of refraction) is shown in Fig. 3.34.

• Cooling-rate and slow-cooling temperature range: when a glass is subjected to constant-rate cooling, the change in index of refraction decreases as the temperature drops, until eventually the index of refraction does not vary at all. If the temperature, where this occurs, is called T_2, Lillie and Ritland[16] consider that $T_2 = T_1 - 90°C$ (T_1 is the soaking temperature). However T_2 is also a function of the cooling rate and can be expressed as shown in Fig. 3.35.

• Relationship between cooling rate and n_T: the relationship between final index of refraction n_T and cooling rate obtained when the glass was cooled at 4.67, 15, 48 and 200°C/hr after soaking at 610°C is shown in Fig. 3.36. The variable n_T is determined by the cooling rate.

Effect of Soaking Temperature. The temperature at which soaking is performed (before constant-rate cooling) has an effect on the final index of refraction. Figure 3.37 shows these soaking effects and depicts the relationship between

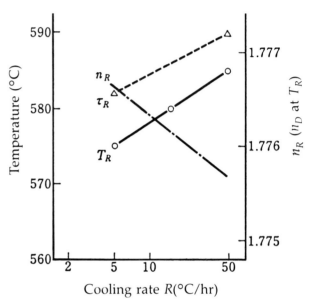

Figure 3.33. Relationship of cooling rate to T_R, τ_R, and n_R (BCL 8).

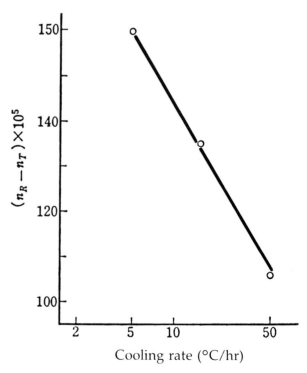

Figure 3.34. Relationship between $n_R - n_T$ and cooling rate (BCL 8).

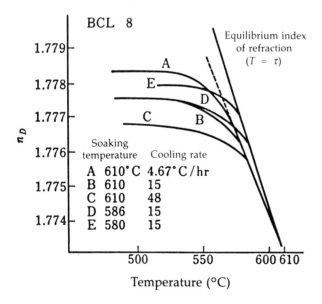

Figure 3.32. Changes in index of refraction during constant-rate cooling.

soaking temperature and n_T. In the case of a fixed cooling rate, if the soaking temperature is higher than τ_R, then n_T is independent of the soaking temperature. If the glass is maintained at a temperature lower than τ_R, the final index of refraction increases, but at the same time the effect of the cooling rate is decreased. In an extreme case, i.e., soaking at 570°C and cooling at 200°C/hr, the index of refraction coincides with the equilibrium index of refraction.

Soak Samples and Rate Samples. Figure 3.38 is Fig. 3.32 redrawn to a larger scale. Point B,

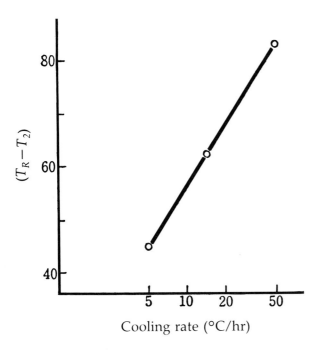

Figure 3.35. Relationship between $T_R - T_2$ and cooling rate (BCL 8).

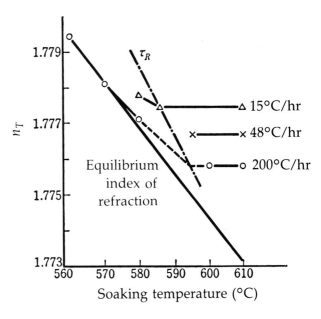

Figure 3.37. Relationship between n_T and soaking temperature (BCL 8).

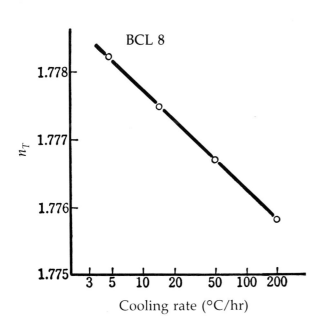

Figure 3.36. Relationship between n_T and cooling rate.

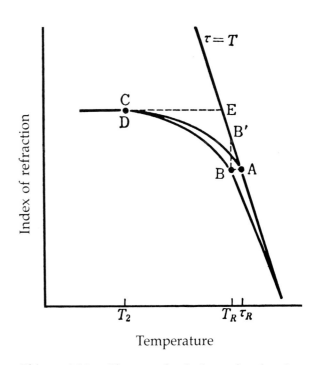

Figure 3.38. Changes in index of refraction during constant-rate cooling.

which is at T_R in the case of a cooling rate at 15°C/hr, has the same structure as point A. Point A shows the same index of refraction on the equilibrium index of refraction line. Furthermore, samples C and D obtained by cooling soak-sample A and rate-sample B at the same rate also show the same index of refraction and appear to have the same structure. The reason for this is as follows: sample B shows absolutely no change even if it is maintained at the equilibrium temperature of A; further, if samples indicated by C and D are maintained at the temperature of E, which shows the same index of refraction as these samples, samples C and D reach the index of refraction of E by exactly the same route. Samples C and D have the same structure; these samples and E show the same index of refraction, but cannot be said to have the same structure. As shown in Fig. 3.39, the index of refraction of C and D first decreases with heat treatment time, then reaches a minimum value and begins to increase, and then finally reaches an equilibrium value E. It appears that in samples C and D, a process with a shorter relaxation time is active down to the temperature T_2. Meanwhile, a process with a longer relaxation time is frozen at higher temperature (A). In the case of sample E, it appears that both a process with a shorter relaxation time and a process with a longer relaxation time are frozen at the temperature of E. Even though soak samples and rate samples may show the same index of refraction, we cannot say that they necessarily have the same structure. In this experiment, furthermore, the fact that we need to consider two relaxation processes with different relaxation times is extremely interesting. The same conclusion was reached regarding changes in index of refraction resulting from constant-temperature soaking.

3.1.3.2.   Changes in Strain Caused by Heat Treatment

In the heat treatment of glasses, more emphasis was originally placed on the relaxation of residual stress than on the adjustment of the index of refraction. The consideration of changes in index of refraction caused by heat treatment is relatively recent; the original purpose of heat-treating optical glasses was to eliminate residual stress. When thermal stress remains in a glass, birefringence occurs. When plane polarized light is passed through a glass containing strain, the birefringence caused by the stress creates a light path difference between ordinary and extraordinary rays. This light path difference is proportional to the magnitude of the stress.

$$S = l \left(n_e - n_0 \right) = l \cdot \beta \cdot F \quad .$$

Here, S is the light path difference (nm) between ordinary rays and extraordinary rays, n_e and n_0 are the respective indices of refraction for extraordinary rays and ordinary rays, l is the length of the light path (cm), β is the stress-optic coefficient (cm²/dyne), and F is the stress.

The quartz-wedge method is widely used for the quantitative measurement of strain at high temperatures. However, the measurement sensitivity of this method is generally limited to about 10 nm/cm, which is insufficient for a detailed study of strain. Here, we will discuss measurement results obtained using the "polarimeter for measuring the reduction of strain at high temperatures" invented by Sakata.[17] This apparatus can measure relatively small light path differences (200 to 0 nm) at any desired temperature with a precision of 0.12 to 0.13 nm and is suited for use in a detailed investigation of the strain reduction process.

An outline of the polarimeter apparatus is shown in Fig. 3.40. A special feature of the apparatus is that it can measure minute variations in the light path difference by means of a combination of an orthogonal sharp color tint plate and a rotary compensator, using brightly colored polarized light. The residual strain of the sample is determined from the rotation angle of the compensator when the color of the measured portion shows no color difference in the colorimetric region of the orthogonal color-tint plate.

$$\varphi_1 = \tan^{-1} \frac{\sin 2\theta \cdot \sin \varphi_2}{1 - \sin^2 2\theta \left(1 - \cos \varphi_2 \right)} \quad ,$$

$$\delta_1 = \lambda_0 \frac{\varphi_1}{2\pi} \quad , \qquad \delta_2 = \lambda_0 \frac{\varphi_2}{2\pi} \quad .$$

Here, φ_1 is the phase difference of the sample, φ_2 is the phase difference of the compensator, θ is the rotation angle of the compensator, δ_1 is the light path difference of the sample, δ_2 is the light path difference of the compensator, and λ_0 is the dark interference wavelength of the color-tint plate.

Relaxation of Strain by Constant-Temperature Soaking.   The relaxation of strain by constant-temperature soaking can be described by the

equation of Adams and Williamson[18] or by Maxwell's equation.

Adams' and Williamson's equation

$$\frac{1}{S} - \frac{1}{S_0} = At \quad .\tag{3.12}$$

Maxwell's equation

$$-\frac{dS}{dt} = \frac{G}{\eta} \cdot S \quad .\tag{3.13}$$

The initial stage of strain relaxation that occurred when samples of the lanthanum glass BCL 8 (composition by weight: 26% B_2O_3, 36% CdO, 27% La_2O_3, 5% SiO_2, 4% ZrO_2, and 2% TiO_2; T_g: 578°C; 40 × 40 × 10 mm plates) were treated at 560, 570, 580, and 590°C was described relatively well by the equation of Adams and Williamson; as time passed, however, the strain relaxation curves were observed to diverge from the predicted values. On the other hand, it was found that Maxwell's equation was able to describe strain relaxation throughout more or less the entire process, except for the initial stage (Fig. 3.41).

If Tool's equation is rewritten as an equation that expresses the stress relaxation rate and then is applied with a fixed soaking temperature to the initial stage (which cannot be described by Maxwell's equation),

$$\ln \frac{dS/dt}{S} = \ln k + \frac{S}{h} \quad .\tag{3.14}$$

It was found that the initial state of stress relaxation is more or less described by this equation

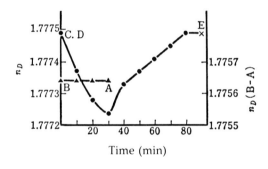

Figure 3.39. Index of refraction of soak samples and rate samples in the case of constant-temperature soaking. Points A and E are soak samples; points B, C, and D are rate samples.

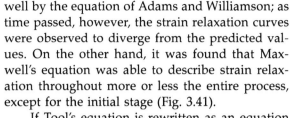

Figure 3.40. Polarimeter for measuring the reduction of strain at high temperatures: (1) light source, (2) condenser lens, (3) ground-glass scattering plate, (4) polarizer, (5) electric furnace, (6) sample, (7) orthogonal sharp-color tint plate, (8) rotary compensator, (9) analyzer, and (10) observer.

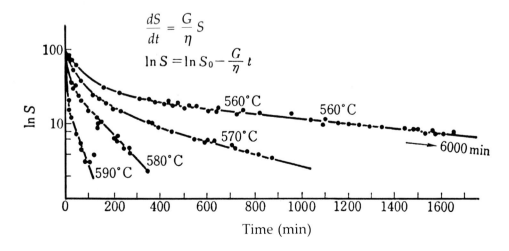

Figure 3.41. Strain relaxation curves (BCL 8).

(Fig. 3.42). The activation energy for the initial stage was 113 kcal/mol and for the second stage was 185 kcal/mol. The activation energy of stress relaxation in the silicate glass LF 6 was 69 kcal/mol.

Generation of Strain in the Cooling Process.

Effect of Soaking Temperature on the Amount of Residual Strain. Samples with residual strain were soaked at various temperatures in the annealing region; after the strain had been completely eliminated, the samples were cooled to room temperature at constant cooling rates. The final amounts of strain that resulted are shown in Fig. 3.43. As seen in Fig. 3.43, a considerable amount of strain is induced when a sample is soaked at a temperature higher than the transition point (578°C) and then cooled. The reason for this is that the expansion coefficient shows an abrupt increase with the transition point as a boundary. Below the transition point, however, even if a glass from which strain has been eliminated is cooled at a fairly rapid rate, the final amount of strain obtained is small.

Relationship between Cooling Rate and Amount of Residual Strain. Samples of BCL 8 were soaked at 580°C; after the strain had been completely eliminated, the samples were cooled at various cooling rates. The amounts of strain that resulted are plotted against cooling rate in Fig. 3.44.

According to Lillie and Ritland,[16] birefringence (A) in the central plane of the sample is given by the following equation:

$$A = \frac{2B}{3} \cdot \frac{E\alpha'}{1 - \sigma} \cdot \frac{a^2 R}{8k} \quad . \qquad (3.15)$$

Here, B is the stress-optic coefficient, E is Young's modulus, σ is Poisson's ratio, α' is the coefficient of thermal expansion in the vicinity of T_g, a is the thickness of the sample, R is the cooling rate, and k is the thermal diffusivity. This equation shows that a linear relationship exists between the amount of residual strain and the cooling rate in the case of a sample of fixed thickness. In the case of BCL 8, assuming that the permissible amount of residual strain is 5 nm/cm, a value of $R = 15$°C/hr is obtained. The stress-optic coefficient B of BCL 8 was 2.35 cm^2/dyne. Furthermore, in Fig. 3.44, a cooling rate of 15°C/hr showed a residual strain of 8.7 nm for a light path length of 40 mm.

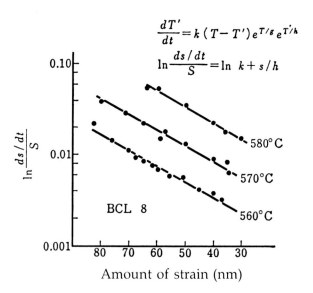

$$\frac{dT'}{dt} = k\,(T - T')\,e^{T/s}\,e^{T'/h}$$

$$\ln \frac{ds/dt}{S} = \ln\ k + s/h$$

Figure 3.42. Initial stages of strain relaxation curves (according to Tool's equation).

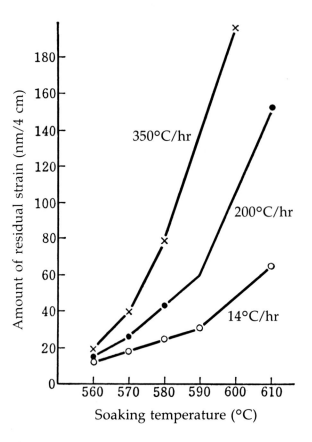

Figure 3.43. Effects of soaking temperature and cooling rate on amount of residual strain.

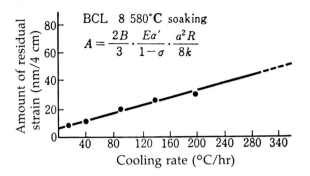

Figure 3.44. **Relationship between amount of residual strain and cooling rate.**

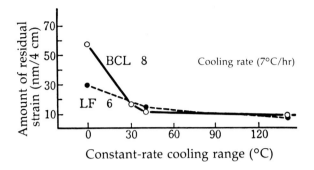

Figure 3.45. **Relationship between amount of residual strain and constant-rate cooling range.**

Slow-Cooling Temperature Range. Samples of BCL 8 were soaked at 580°C; after the strain had been completely eliminated, the samples were slow cooled to temperatures of 550, 540, 500, and 440°C at a rate of 7°C/hr and were then rapidly cooled. The amounts of residual strain that resulted are plotted against the temperature range to the lower-limit temperature of slow cooling in Fig. 3.45. The figure also shows results for LF 6. As Fig. 3.45 shows, an interval of approximately 40°C below the transition point is required for BCL 8, and an interval of more than 50°C below the transition point is required for LF 6, in order to reach a more or less fixed amount of strain. Lillie and Ritland[16] consider the interval between the soaking temperature T_1 and the lower-limit temperature of slow-cooling T_2 to be approximately 90°C.

Stress Relaxation Mechanisms. The relaxation of thermal stress in glasses is thought to consist of at least two relaxation processes, and may be expressed as follows:

$$S = S_1 \exp\left(\frac{-t}{\tau_1}\right) + S_2 \exp\left(\frac{-t}{\tau_2}\right) \quad .$$

The activation energies of the respective terms are 107 kcal/mol and 185 kcal/mol for BCL 8. It ap-

pears that the latter term corresponds to viscous flow, whereas the former term correponds to delayed elasticity. Expressed by Tool's equation,

$$\frac{dT'}{dt} = K(T - T') \exp\left(\frac{T}{g}\right) \cdot \exp\left(\frac{T'}{h}\right) \quad .$$

The flow term is expressed as

$$K \exp\left(\frac{T}{g}\right) \cdot \exp\left(\frac{T'}{h}\right) \quad ,$$

and is influenced not only by ordinary temperatures, but also by the structural temperature. When the difference between the treatment temperature and the structural temperature is great, Tool's equation, which covers changes in the flow term according to structural temperature, is appropriate. When the structural temperature approaches the treatment temperature, the flow term becomes fixed, and Tool's equation turns into Maxwell's equation:

$$\frac{dT'}{dt} = K_T(T - T') \quad .$$

It appears that this is the reason Maxwell's equation applies so well to the later stage of strain relaxation.

3.2. Progress in Optical Glass Manufacturing Methods

Optical glass manufacturing methods underwent a great change in 1965. That is to say, in 1965, with the initiation of continuous melting of BK 7 at Hoya Glass, optical glass manufacturing

methods underwent a great conversion from intermittent production systems to continuous production systems. The optical glass manufacturing methods in use previous to 1965 were methods introduced from the German firm of Schott, Ltd., and were established in Japan through the incredible efforts of Dr. Toru Takamatsu at the Osaka Industrial Research Institute and Mr. Masao Nagaoka at Nippon Kogaku. With the industrialization of new lanthanum borate glasses in 1945, a platinum pot melting method was developed at Eastman Kodak Company in the United States. Platinum pot melting of optical glasses was also begun in Japan. With this method, there is no pot erosion of the kind seen in clay pot melting, and melting faults such as bubbles, striae, and coloring are reduced. Thus, the development of this method represented a great step forward.

In 1965, a continuous manufacturing technique was developed, and a system was established in which binocular prisms were melted, pressed, and annealed in a continuous process. However, glasses for use in photography were manufactured in the form of strips or E-bars (extruded bars), formed into thick plates (30 to 50 mm) or rods, and hand-pressed after cutting. Photographic lens blanks began to be produced in 1971 using a 3-D system, i.e., direct melting, direct pressing, and direct annealing.

Table 3.10 shows a comparison of optical glass manufacturing systems. As shown by the bottom rows in the table, great progress has been achieved in moving from clay pot melting to the platinum-tank continuous-production system: the number of days required for production has been reduced from 170 days to 3 days, and the yield has been increased from 40% in just the melting stage to a pressed product yield of 90%.

3.2.1. Clay Pot Melting

Clay pot melting was the first system used in melting optical glasses. Because the pots were made of clay, the glass melts caused erosion. The manufacture of corrosion-resistant pots was a prime task; even in the case of SK glass melts containing Ba, the prevention of glass leakage was not easy. The appearance of lanthanum borate glasses led inevitably to a switch to platinum pots. Because the pots were pulled out of the furnace after melting, resistance to sporing was important. In addition, because the pots themselves had a porosity, bubbles were easily generated from the clay pot wall as a result of incomplete glass wetting, and coloring of the glass by iron contained in the clay was unavoidable. The greatest problem, however, was striae; the pots themselves were eroded and dissolved, leading to striae. Accord-

ingly, even if the mixing of the melt was promoted by stirring, the generation of striae was unavoidable. Stirring was performed more with the idea of fixing striae in the wall areas and central areas of the pots, and thus preventing their spread throughout the melt, than with any idea of eliminating them. Therefore, it was difficult for the melting yield to exceed 40%. Studies on clay pots used for optical glasses include the work of Ono.[19]

3.2.2. Platinum Pot Melting

Lanthanum borate glasses have a low viscosity (several poise to several tens of poise) and are like water compared with silicate glasses; accordingly, the use of platinum pots was unavoidable.

In the case of platinum pots, the wetting of the glass is good and there are no pores; consequently, there is little bubble generation. Because erosion can also be ignored, there are no striae generated in this manner. As a result, this method has increased the melting yield to 80%. Furthermore, coloring of the glass by iron was also eliminated. In glasses containing large amounts of lead, however, coloring by lead platinate was unavoidable.

In platinum-pot melting, unlike clay-pot melting, the stirring speed could be raised without entraining bubbles; as a result, stirring for the purpose of eliminating striae became possible. The weakest point of this sytem was as follows: because platinum pots were used, mass production was impossible; production was limited to multipot or multistage production systems. We might say that this system was suitable for the production of small amounts of a variety of glasses. Accordingly, the production efficiency of this manufacturing method were poor. Various heating systems were used, e.g., furnaces with silicon carbide heating elements, high-frequency induction heating, etc., but these did not cause any essential change in the system.

3.2.3. Continuous Melting

Around 1960, continuous melting of glass for use in eyeglasses was performed at Bausch & Lomb and Corning in the United States. However, serious doubts existed as to whether these systems could be applied to the production of optical glasses in which higher quality was required.

In 1960, at the Osaka Industrial Experimental Institute, Izumitani et al. attempted to develop

Table 3.10. Comparison of optical glass manufacturing systems.

Clay pot melting Intermittent		Platinum pot melting Intermittent		Platinum tank melting Continuous			
Ingot		Cast form		Strip		Direct pressing	
Pot manufacture	120 days	Metal pot		Platinum tank		Platinum tank	
Preheating	5 days						
Firing	1 day						
Melting	1 day	Melting ⎫ Casting ⎬	1/2 day	Melting ⎫ Forming (strips) ⎬	1 day	Melting ⎫	
Cooling	7 days	Cooling	5 days	Cooling ⎭			
Pot breaking	1 day	Cutting	1/2 day				
Ingot selection	3 days	Polishing	3 days				2 days
Forming	5 days						
Polishing	3 days						
Inspection and measurement of material	1 day	Inspection and measurement of material	1 day	Inspection and measurement of material	1 day		
Cutting	1 day	Cutting	1 day	Cutting	1 day		
Pressing		Pressing		Pressing		Pressing	
Annealing	21 days	Annealing	21 days	Annealing	21 days	Annealing ⎭	
Inspection and measurement of pressed products	1 day	Inspection and measurement of pressed products	1 day	Inspection and measurement of pressed products	1 day	Inspection and measurement of pressed products	1 day
Pressed products		Pressed products		Pressed products		Pressed products	
Days required for production	170 days	Days required for production	34 days	Days required for production	25 days	Days required for production	3 days
Melting yield	40%	Melting yield	70%	Melting yield	90%	Pressed product yield (maximum)	90%

high-frequency continuous forming of strips, but failed. Around 1963, Izumitani et al., at Hoya attempted the continuous melting of optical glasses; in 1965, they succeeded, and the system was adopted for actual production. The concept of the system is described in "An Apparatus for the Continuous Manufacture of Homogeneous Optical Glass" (Patent Application Publication No. 43-12885).

3.2.3.1. Principles of Continuous Melting

Izumitani's concept is based on the following three points:

1. The passage of time in pot melting is replaced by a spatial arrangement.

2. The system is devised so that the glass flows in one direction; flow in the opposite direction is prevented by pipe connections.

3. The melting apparatus consists of several chambers; functions are performed one at a time in these chambers. Specifically, the apparatus consists of a melting vessel, fining vessel, stirring vessel, operating vessel, etc. Melting, bubble re-moval, elimination of striae, and adjustment of the forming viscosity are performed in the respective vessels.

A continuous tank furnace was built based on the above structural concept. In this furnace, raw materials are fed in at one end, and glass, which is free of bubbles and striae, flows out of the operating zone. Figure 3.46 shows one example of this system. Depending on the type of glass involved, (1) and (2), or (4) and (5) may be combined, or (2) may be omitted.

In clay pot melting, the following process is performed:

1. Feeding of raw materials,

2. Stirring to mix the raw materials and remove bubbles,

3. Allowing the melt to stand to remove bubbles (fining),

4. Lowering the temperature,

5. Stirring to eliminate striae,

6. Removing of the pot.

The above procedure allows the manufacture of a homogeneous glass with the elimination of

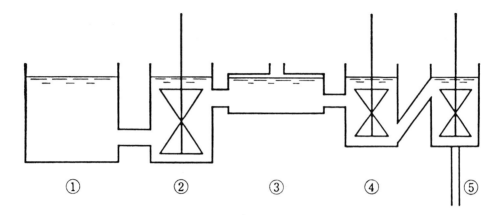

Figure 3.46.   Model diagram of continuous tank furnace for optical glasses.

striae and bubbles. If this system is performed spatially, it should certainly allow the manufacture of a homogeneous glass. This was Izumitani's idea. Accordingly, several vessels were installed and connected by pipes. The drawback of conventional plate or bottle glass tanks is that they generate reverse flow. This has led to the following drawbacks: batches run too far ahead, glass that has undergone insufficient fining runs too far ahead, and glass that has been homogenized flows backward. There has also been another drawback: the more the glass is homogenized, the greater the concentration of heterogeneous portions in certain areas. In the present system, therefore, the vessels were connected by pipes to allow each vessel to perform its function; connecting the vessels permitted the homogenized glass to flow in a successive laminar flow, and ensured that heterogeneous glass did not run too far ahead. In each tank, precisely the function of that tank was performed. The system was designed so that bubbles were completely eliminated in the fining tank and so that striae were completely eliminated in the stirring vessel. This concept was attended by success; in 1964, bubble-free, striae-free BK 7 glass flowed out of the system. As an engineer, this was an unforgettable moment of deep emotion for the author.

The continuous melting of optical glasses is based on the principle of "utilizing the special features of continuous tank-furnace melting and intermittent pot melting, creating an overall piston flow while achieving complete mixing and continuously extracting a homogeneous glass." Naturally, the structure of the system is changed according to the type of optical glass involved. As a rule, however, the minimal tank structure required consists of (1) a melting vessel equipped with a stirring apparatus for the purpose of mixing

the raw materials, (2) a fining vessel for the purpose of bubble removal, and (3) a conditioning vessel equipped with a stirring apparatus, which eliminates striae and adjusts the temperature for forming. This continuously melted glass is combined with automatic pressing apparatus and automatic annealing apparatus; as a result, we have now reached a point where completely pressed products can be removed from the output end of the production line within one or two days after the raw materials have been inserted, with the yield in terms of pressed products exceeding 90%.

3.2.4.   Electric Melting

For the most part, propane gas was used to heat optical-glass melting tanks. From around 1975, however, attempts have been made to melt glass by the direct application of electric current. The reason for such attempts is as follows: in the case of a small furnace, gas combustion is difficult and achieves only a poor heating efficiency. Electric current melting, on the other hand, generates little pollution and is able to heat a glass to high temperatures even in a small melting furnace.

3.2.4.1.   Electric Resistance of Optical Glasses

Electric current melting utilizes the joule heat generated by the electric resistance of the glass melt. Figure 3.47 shows the relationship between resistivity and temperature for representative optical glasses and refractories.

3.2.4.2.   Electrode Materials

Molydenum (Mo) and SnO_2 are used as electrode materials. The material Mo has a good electric conductivity and is resistant to thermal shock.

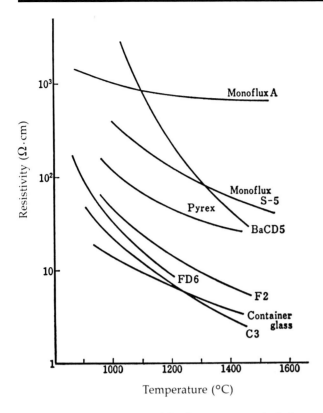

Figure 3.47.   Relationship between resistivity and temperature for representative glasses and refractories.

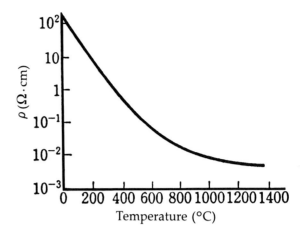

Figure 3.48.   Example of resistivity of a tin oxide electrode.

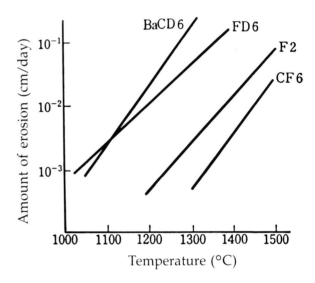

Figure 3.49.   Relationship between temperature and amount of erosion of a tin oxide electrode.

When heated in air, however, it oxidizes and volatilizes; accordingly, water cooling is necessary. Furthermore, Mo cannot be used in melts of glasses that contain easily reduced oxides, e.g., PbO or Sb_2O_3. Because As_2O_3 and Sb_2O_3 are commonly used as refining (bubble removing) agents in the case of optical glasses, the use of Mo is limited.

Because tin oxide does not reduce the glass, it can also be used in melts of glasses containing PbO. In most electrodes, small amounts of Sb_2O_3 are added to decrease the electrical resistance, and small amounts of Cu are added to accelerate the sintering rate. Figure 3.48 shows the relationship between resistivity and temperature in a tin oxide electrode. Drawbacks of tin oxide electrodes are that they have a poor resistance to thermal shock, and that they are eroded by the glass at high temperatures. Figure 3.49 shows the relationship between temperature and the amount of erosion of a tin oxide electrode by various optical glasses. The maximum electrode current density is 0.7 A/cm^2.

3.2.4.3.   Furnace Structure

In an electric-current melting furnace, the surface of the glass melt is covered with a batch.

The temperature reaches a maximum in the vicinity of the electrodes and drops as the glass moves downward. In other words, a special feature of such a furnace is that the melting of the raw materials, fining, and squeezing* are performed in a vertical direction (unlike the case of a horizontal type tank). A semimolten layer is present at the boundary between the batch layer and the glass melt. This layer is thought to consist of unmelted raw materials, eutectically melted liquid and gas bubbles. This layer is in balance with the upward convective flow and is moved downward in piston form.

---

* Shrinkage of small bubbles by gas dissolution into the melt.

A vertical-type single-tank furnace is useful from the standpoint of saving energy. However, there is a tendency for the batch to run too far ahead; homogenization is insufficient, and the index of refraction of the outflowing glass tends to fluctuate. If a horizontal-type two-tank structure is used in such a case, the function of melting and fining can be separated. Homogenization is promoted by the two-tank structure, and there is little fluctuation in the index of refraction of the outflowing glass. In this case, the mixing and melting of the batch can be effectively promoted by bubbling or stirring in the melting tank. The melting capability of such a small electric current furnace is 0.4 to 0.55 m^2/ton/day. A special feature of this type of furnace is that the thermal efficiency is around 20% and is thus high compared with the 3 to 4% efficiency of gas combustion.

3.2.5. Lenses Formed by High-Precision Pressing

The surface precision of conventional pressed glass products is 1/10 mm. On the other hand, the surface precision of pressed metal is 1/100 mm. Why can't the same pressing precision that is obtained in metal be obtained in glass? The answer lies in the shrinkage that is caused by the temperature difference between the internal part of the glass and the surface. Even if the glass is formed during pressing so that the same shape as the metal mold is obtained, the surface temperature is lowered by pressing so that the surface

layer solidifies, while the internal part of the glass remains in a liquid state at high temperature. As the glass is cooled to ordinary temperatures, the temperature of the internal part of the glass also drops; the internal part solidifies and contracts so that the surface shape changes, and shrinkage is produced.

Kitayama and Izumitani[20] found that the surface precision can be controlled to within 1/100 mm by pressing for 20 seconds or longer at a temperature between Sp and Tg. They succeeded in obtaining a high surface precision by forming at a temperature between Sp and Tg, by lowering the temperature difference between the surface and the internal part of the formed glass by maintaining the glass at that temperature for a fixed period of time, and then by cooling the glass "as is."

Recently (in 1983), spherical and nonspherical pressed lenses were announced by Kodak. This may be called a revolutionary new technique that replaces polishing and surpasses polishing in the case of nonspherical lenses. However, the principles of pressed lens manufacture are based on the high-precision pressing techniques described above. Preformed blanks that are free of share marks are used; metal-mold materials that will not oxidize at high temperatures and cannot be wetted by the glass are selected and worked to have mirror surfaces. By controlling the atmosphere, the pressing temperature, the pressing time, and the pressing pressure, the mirror surfaces of the metal mold can be transferred to the glass. Figure 3.50 shows interference fringes of the metal-mold surfaces and the pressed-lens surfaces. Surfaces

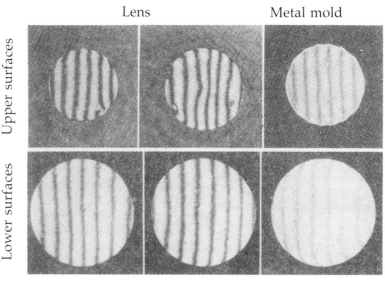

Figure 3.50. Interference fringes of metal-mold surfaces and pressed-lens surfaces.

with a surface roughness of 100 Å, a surface precision of λ, and an astigmatism of $\lambda/4$ were obtained.

Lens manufacture by pressing differs from lens manufacture by polishing in that there is no need to consider staining, hardness, etc., of the glass; consequently, the glass composition can be selected with greater freedom. It appears that such advantages will increase, and that the appearance of pressed lenses will not only facilitate the manufacture of aspherical lenses, but also lead to a great revolution in the world of optics.

References

1. T. Izumitani et al., *Rep. Osaka Ind. Res. Inst.*, No. 319, 78 (1962).
2. H. Jebsen-Marwedel, *Gastechn. Ber.* **29**, 233 (1956)
3. T. Izumitani et al., *Rep. Osaka Ind. Res. Inst.*, No. 319, 88 (1962).
4. S. Nagata et al., *Chem. Eng. Japan* **21**, 278 (1957).
5. M. Volmer, *Kinetik der Phasenbildung* (Dresden, Steikopff., 1939).
6. M. Cable, *Glass Techn.* **2**, 60 (1961).
7. T. Izumitani et al., *Rep. Osaka Ind. Res. Inst.*, No. 319, 108 (1962).
8. C. L. Babcock and D. A. McGraw, *Glass Ind.* **38**, 137 (1957); W. Trier, *J. Am. Ceram. Soc.* **44**, 339 (1961).
9. M. Coenen, *Adv. Glass Technology*, New York, 93 (1962).
10. A. Q. Tool, *J. Am. Ceram. Soc.* **29**, 240 (1946).
11. T. Izumitani et al., *Rep. Osaka Ind. Res. Inst.*, No. 319, 125 (1962).
12. A. Winter, *J. Am. Ceram. Soc.* **26**, 189 (1943).
13. H. A. McMaster, *J. Am. Ceram. Soc.* **28**, 1 (1945).
14. P. W. Collyer, *J Am. Ceram. Soc.* **30**, 338 (1947).
15. H. N. Ritland, *J. Am. Ceram. Soc.* **37**, 370 (1954).
16. H. R. Lillie and H. N. Ritland, *J. Am. Ceram. Soc.* **37**, 466 (1954).
17. H. Sakata and Kitaoka, *Rep. Osaka Ind. Res. Inst.*, No. 9, 161 (1958).
18. L. H. Adams and E. D. Williamson, *J. Franklin Ins.* **190**, 157 (1920).
19. Ono, *Yogyo-Kyokai-shi* **63**, 565 (1955).
20. Kitayama, T. Izumitani, and Ogawa, Japanese Patent Application No. 46-19002.

CHAPTER 4

Cold Working of Optical Glasses

After being pressed or cut, optical glasses are ground and polished into lenses or prisms. The cold working of lenses and prisms can be divided into three processes: roughing (rough grinding), smoothing (fine grinding), and polishing. In the first two processes, glass is removed by mechanical fracture. In the polishing process, glass is removed through plastic flow caused by scratching (abrasion). Problems that arise in the glass polishing process are dimming, staining, and latent abrasion. The latter problem occurs in the washing process of lenses. Furthermore, the glass washing process has an intimate relationship with the coating of the glass surface. The adhesive strength of a coating is affected by surface contamination, dimming, and staining, and depends on the composition of the glass. In this chapter, the processes of polishing, grinding, washing, and coating are treated with respect to the relationship of these processes to the physical properties of glasses; in addition, the problems of dimming, staining, and latent scratching in the polishing and washing processes are also discussed.

4.1. The Polishing Process

The cold working of glasses is performed in the following order: diamond grinding, lapping, and polishing. Here, however, we will discuss these processes in reverse order: i.e., polishing, lapping, and diamond grinding.

Polishing is usually performed using a polisher such as a pitch or a polyurethane-foam pad, etc., and a polishing agent such as cerium oxide, etc. Here, we will discuss the physical properties and functions of polishing agents and polishers.

4.1.1. The Mechanism of Polishing

How is a mirror surface obtained in glass? How are irregularities in the glass surface eliminated? There are three theories concerning the mechanism of glass polishing:

1. The mechanical microscopic cutting theory (Refs. 1–3),
2. The flow theory (Refs. 1, 2, 4–6),
3. The chemical reaction theory (Refs. 1, 2, 7).

The mechanical microscopic-cutting theory was proposed by Rayleigh. According to this theory, the polishing of the glass surface is accomplished by mechanical fracture and removal on a scale on the order of molecules. The flow theory was proposed by Beilby. According to this theory, the surface of the glass is locally and instantaneously heated by friction with the polishing agent. As a result, the protruding parts of the surface are caused to flow into the valley areas by viscous flow so that the irregularity of the surface is filled in. The chemical theory was proposed by Preston, and then by Grebenshchikov. The latter proposed that a silica gel layer is formed on the surface of the glass by a reaction with water, and that this gel layer is removed by the polishing agent. Which of these theories is correct?

Brüche et al., (Ref. 8) studied the mechanism of glass polishing from the standpoint of the physical properties of glasses. Specifically, if the microscopic cutting theory is correct, then the glass polishing rate should depend on the hardness or strength of the glass. If the flow theory is correct, then the polishing rate should depend on the softening point of the glass. Finally, if the chemical theory is correct, then the polishing rate should depend on the chemical durability of the glass. Furthermore, the polishing theory must be able to explain the functions of polishing agents, polishing liquids, and polishers without contradiction.

4.1.1.1. Relationship between Polishing Rate and Physical-Chemical Properties of Glasses

Approximately 200 types of optical glasses exist; from these, 18 types of glass with different hardnesses, softening points, and chemical durabilities were selected. Table 4.1 shows various properties of these glasses. In Table 4.1 the softening point was determined from the yield point of the thermal expansion curve. The Vickers hardness was measured by applying a load of 25 g to the polished surface for 15 seconds. Finally, the chemical durability was measured by immersing glass particles (420 to 590 μm) in (a) a 0.01 N solution of nitric acid and (b) water for one hour at 100°C, and measuring the weight loss.

The glass samples were lapped with aluminum oxide (mean particle size: 18 μm) and were then polished with an Oskar polisher using polyurethane pads and a polishing slurry prepared by adding 10 g of cerium oxide to 100 ml of water. The polishing rate was determined from the relationship between the mean surface roughness (PVA) and the polishing time.

Figure 4.1 shows the relationship between the polishing rate and the micro-Vickers hardness of the glass. No correlation is observed between the two; this indicates that the glass is not polished by microscopic cutting. Figure 4.2 shows the relationship between polishing rate and softening point. Here, too, no correlation is observed between the two. Accordingly, the flow theory is incorrect. Figure 4.3 shows the relationship between polishing rate and acid-resistance weight loss. Here, respective linear relationships between the two values are seen for silicate glasses and borate glasses. As shown in Figure 4.4, an approximately proportional relationship is observed between polishing rate and water-resistance weight loss. Thus, the polishing rate of a glass depends on the chemical durability of the glass. In other words, a chemical

reaction between the glass and the polishing liquid plays an important role in the polishing process.

As shown in Fig. 4.3, however, two different linear relationships are involved: one in the case of silicate glasses, and one in the case of borate glasses. This indicates that the polishing rate is determined by chemical durability and by another factor. What is the significance of the fact that the polishing rate of a glass is determined primarily by the chemical durability of the glass?

4.1.1.2. Formation of a Hydrated Layer

Polishing (which is performed mechanically) and chemical durability (which is governed by chemical reactions) appear to be widely separated properties. However, both of these properties affect the glass surface. It appears that some kind of change occurs in the glass surface during polishing, and that this change is related to polishing.

To ascertain changes in the surface, the hardness of the glass surface layer was measured. Figure 4.5 shows the relationship between the load applied to the Vickers indenter and the Vickers hardness of the glass. In a freshly polished surface, the micro-Vickers hardness ordinarily increases as the load is decreased. However, the Vickers hardness of a glass surface that has been immersed for 60 minutes in a 0.1 N solution of HCl decreases with a decrease in the load.

Table 4.1.   Physical–chemical properties of glass.

Glass type	Softening point (°C)	Vickers hardness (kg/mm^2)	Acid-resistance weight loss (%)	Water-resistance weight loss (%)
BK 7	615	707	0.08	0.13
KF 2	490	627	0.07	0.07
SK 2	700	707	0.70	0.05
SK 16	680	689	3.3	0.58
SF 6	470	413	1.3	0.03
FK 1	475	666	1.9	—
SF 13	480	437	0.34	0.02
KF 3	500	627	0.04	0.05
BaF 4	580	613	0.11	0.04
BaK 4	620	657	0.43	—
F 3	480	548	—	—
LaKF 2	675	803	1.3	0.25
LaK 12	670	743	1.7	0.35
LaLK 13	650	762	1.9	0.70
NbF 1	650	824	1.0	0.01
TaF 2	685	847	0.74	0.01
LaK 10	670	803	1.2	0.02
NbSF 3	650	803	0.76	0.01

This indicates that a soft hydrated layer is formed on the glass surface.

Figure 4.6 shows the relationship between immersion time and hardness of the surface layer observed when the glass samples were immersed in a 0.1 N solution of HCl. As shown by the figure, the hardness of the surface decreases as the immersion time becomes longer; this indicates that a soft layer is formed by a chemical reaction. This soft layer is thought to be a hydrated layer. The formation of an SiO_2-rich hydrated layer on the surface of a glass as a result of an ion-exchange reaction between modifier ions in the glass and hydrogen ions in the water is a well-known phenomenon. It is thought that a hydrated layer is more easily formed in glasses with poorer chemical durability.

Figure 4.7 shows the relationship between the Vickers hardness of the hydrated layer and the

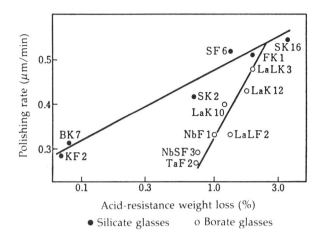

Figure 4.3. Relationship between polishing rate and acid resistance of the glass.

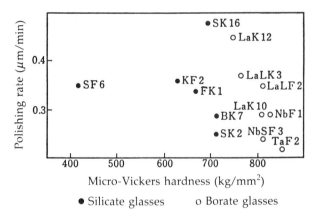

Figure 4.1. Relationship between polishing rate and micro-Vickers hardness of the glass.

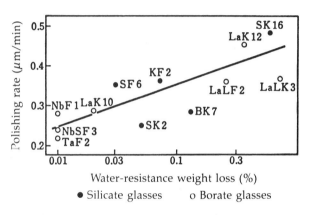

Figure 4.4. Relationship between polishing rate and water resistance of the glass.

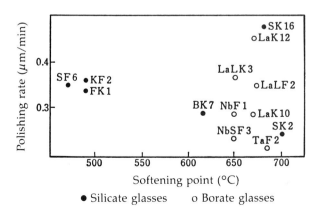

Figure 4.2. Relationship between polishing rate and softening point of the glass.

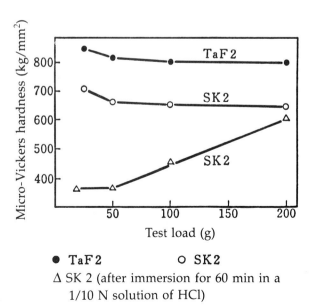

Figure 4.5. Relationship between test load and micro-Vickers hardness of the glass.

polishing rate. The polishing rate is inversely proportional to the hardness of the hydrated layer. As the hardness of the hydrated layer decreases, the polishing rate increases. Thus, it was found that another factor determining the polishing rate is the hardness of the hydrated layer. Therefore, it is clear that the polishing rate increases with an increase in the ease with which a hydrated layer is formed (decrease in chemical durability) and with a decrease in the hardness of the hydrated layer. The difference in polishing rate between lanthanum borate glasses and silicate glasses having the same hydrated layer hardness can be attributed to a difference in the rate of formation of the hydrated layer.

4.1.1.3. Polishing Mechanism

An electron microscope and a differential interference microscope were used to determine how a lapped surface is turned into a smooth surface by the polishing process, and to determine what kind of effect the polishing abrasive particles have on the glass surface. When observed with an electron microscope, an SF 6 surface that had been polished with a cerium oxide polishing agent showed no irregularities. However, when the same surface was immersed in nitric acid for 15 seconds, numerous scratches appeared. When the surface was polished with harder aluminum oxide abrasive particles, the scratches appeared more clearly. These electron micrographs (Fig. 4.8a, b, and c) indicate that the glass surface is scratched by the polishing abrasive particles.

The process by which smooth mirror surfaces are obtained in SSK 4 was observed with a differential interference microscope (Fig. 4.9). The polished surface was immersed in 0.1 N hydrofluoric acid for one minute and was then sputtered with chromium. The lapped surface consisted of conchoidal cracks; as polishing progressed, however, scratches appeared, and the fractured surface area decreased. Ultimately, a polished surface covered with scratches appeared. These photographs indicate that the polishing abrasive particles form a smooth surface by scratching the rough glass surface.

Next, the weight loss occurring during polishing was measured. Figure 4.10 shows the relationship of polishing time to surface roughness and weight loss of the glass. A clear weight loss is observed during polishing. Figure 4.10 shows that the surface roughness decreases as weight is lost, i.e., that polishing proceeds by the removal of material.

4.1.1.4. Summary

The fact that surfaces obtained by etching polished surfaces are entirely covered by

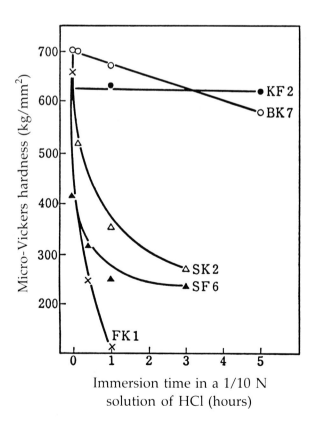

Figure 4.6. Relationship between micro-Vickers hardness and immersion time in a solution of hydrochloric acid.

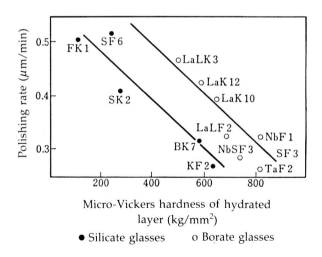

Figure 4.7. Relationship between polishing rate and micro-Vickers hardness of the hydrated layer.

scratches, and the fact that polishing is accompanied by a weight loss, indicate the polishing proceeds by the removal of material. It has been seen that when water is used in polishing, a soft hydrated layer is formed on the glass surface, and that a smooth surface is formed as a result of this hydrated layer being removed by the abrasive particles. Accordingly, the polishing rate is determined by the relative ease with which a hydrated layer is formed (i.e., the chemical durability of the glass) and the hardness of the hydrated layer. Thus, polishing proceeds by the formation of a hydrated layer by means of a chemical reaction between the glass surface and water and then the removal of this hydrated layer by the abrasive particles. It has become clear that a glass mirror surface is formed as a result of the hydrated layer produced by a reaction between the glass and the polishing liquid being planed away by the abrasive particles embedded in the polisher.

 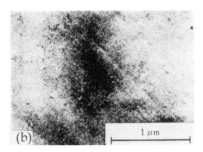 

Figure 4.8. Electron micrographs of glass surfaces: (a) surface polished with CeO_2 plus water, (b) surface obtained by etching the surface shown in (a) with a nitric acid solution, and (c) surface polished with Al_2O_3 plus water.

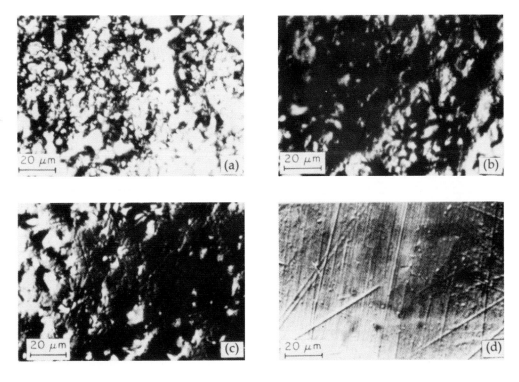

Figure 4.9. Polishing process in SSK 4: (a) lapped surface, (b) surface polished for 2 min, (c) surface polished for 8 min, and (d) surface polished for 14 min.

4.1.2. Polishing Agents (Abrasive Particles)

Cerium oxide is generally used as a polishing agent. Why is cerium oxide used as a polishing agent? In addition to cerium oxide, iron oxide, and chromium oxide have been used in the past, and zirconium oxide has recently come into use. The Mohs hardness of a glass ranges from 5 to 6; accordingly, the polishing agent must have a hardness of 5 to 7. According to Kaller,[10] the polishing rate is proportional to the hardness of the polishing agent, but reaches a maximum at a hardness of 6.5 and then decreases at hardnesses above this value. Accordingly, Kaller believed that the polishing agent must be surface active, and therefore must be a polyvalent oxide. Cornish[11] also surmised that Ce–O–Si bonds are formed between lattice defects in cerium oxide and the surface layer of the glass. Does the polishing agent act through chemical adsorption or mechanical scratching? What in fact is the function of the polishing agent?

4.1.2.1. Function of the Polishing Agent

Figure 4.11 shows changes in surface roughness according to polishing time. When cerium oxide is used in polishing, the roughness decreases with an increase in polishing time; after 15 minutes, a smooth surface is obtained. As shown by Fig. 4.12, however, there is almost no change in surface roughness even after 6 hours when cerium oxide is absent. It is thus seen that a polishing agent must be used to obtain a smooth surface.

Figure 4.13 shows the relationship between the polishing rate and the concentration of the

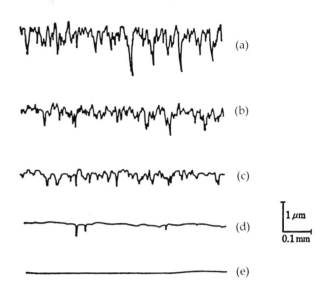

Figure 4.11. Changes in surface roughness in polishing with cerium oxide and water: (a) lapped surface, (b) 3 min of polishing, (c) 6 min of polishing, (d) 10 min of polishing, and (e) 15 min of polishing.

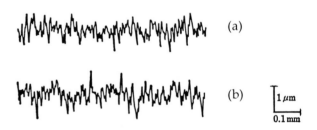

Figure 4.12. Changes in surface roughness in polishing with water only: (a) lapped surface and (b) 6 hr of polishing.

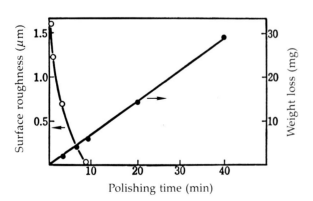

Figure 4.10. Relationship of polishing time to surface roughness and weight loss of glass.

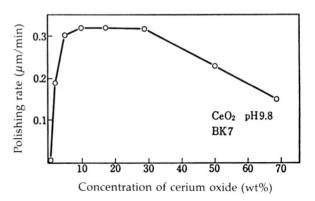

Figure 4.13. Relationship between polishing rate and concentration of polishing agent.

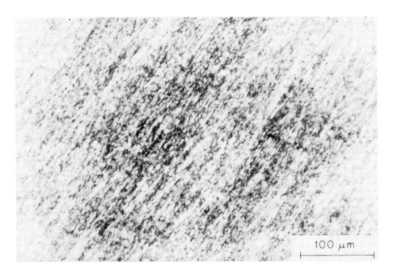

Figure 4.14. Micrograph of a surface polished using cerium oxide and a polyurethane pad.

polishing agent. The polishing rate increases abruptly up to a concentration of 5%, reaches a maximum at 10%, and decreases above 30%. This also indicates that a certain minimum concentration of the polishing agent is required to achieve a corresponding polishing rate, and that a mirror surface cannot be obtained through a chemical reaction with water alone. Therefore, a polishing agent is required to obtain a mirror surface.

Figure 4.14 shows a photograph (taken with a reflecting microscope) of a surface that was polished using cerium oxide and a polyurethane pad and was then coated with chromium by vacuum evaporation. The surface is covered with countless scratches, indicating that the polishing agent acts to scratch the surface of the glass.

4.1.2.2. Relationship between the Polishing Rate and the Hardness of the Polishing Agent

Figure 4.15 shows the relationship of the sintering temperature of cerium oxide to the polishing rate and density. The polishing rate increases with the sintering temperature up to 900°C, where a maximum is reached, and then decreases. Meanwhile, the density increases with the sintering temperature, indicating that sintering is proceeding. Furthermore, it was confirmed by means of x rays that the cerium oxide sintered during this period is tetravalent cerium, with no trivalent cerium present.

Figure 4.16a and b shows changes in particle form over time in a case where samples of cerium oxide sintered at 1400°C and 900°C were crushed

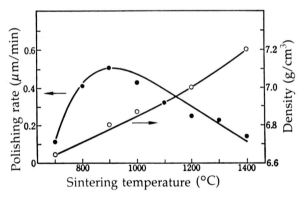

Figure 4.15. Relationship of sintering temperature of cerium oxide to density and polishing rate.

between two glass plates by the application of a pressure of 100 g/cm². The sample sintered at 1400°C showed almost no crushing, whereas the sample sintered at 900°C was finely crushed. Furthermore, as shown in Table 4.2, finer particles result in a higher polishing rate. The reason for this appears to be as follows: easily-crushed abrasive particles that have been fired at a low temperature are crushed into finer, more numerous particles during polishing; as a result, the particles scratch the surface with greater frequency and produce a higher polishing rate. Consequently, a maximum polishing rate is obtained when the cerium oxide has an optimal hardness, i.e., a hardness that is soft enough to allow sufficient crushing during polishing, but which is greater than the

Figure 4.16(a). Changes in the particle form of cerium oxide sintered at 1400°C and then crushed between two glass plates by the application of 100g/cm² pressure.

Figure 4.16(b). Changes in the particle form of cerium oxide sintered at 900°C and then crushed between two glass plates by the application of 100g/cm² pressure.

Table 4.2. Polishing rates obtained with cerium oxide sintered at 900°C and 1400°C.

Sintering termperature	Grain size (μm)	
(°C)	50 to 300	<50
900	0.08	0.39
1400	0.05	0.23

hardness of the glass surface. In other words, the function of the polishing agent is to scratch away the hydrated layer on the glass surface. The mechanism of removal appears to be mechanical scratching rather than chemical adsorption[12]; accordingly, it appears that the polishing agent must have an appropriate hardness, i.e., one that allows sufficient crushing during polishing, but which is greater than the hardness of the glass surface.

4.1.3. Polishing Materials (Polishers)

Glasses are polished using abrasive particles of cerium oxide, etc., and a polisher such as a pitch polisher or polyurethane foam pad, etc. What is the action of pitch? Wada and Hirose[13] measured the rheological properties of pitch. In this section, we will discuss the function of the polisher based on the viscoelastic properties of various polishers, and will describe the relationship of the viscoelastic properties of the polisher to the polishing rate, surface conditions (texture) of the polished surface, and geometrical surface precision.

4.1.3.1. Function of the Polisher

Under what polishing conditions is a mirror surface obtained? To answer this question, tests were performed using various combinations of polishers and abrasive particles. Cast iron polishers (rigid body) and polyurethane pads (viscoelastic body) were selected as polishers, and alumina (Mohs hardness of 9) and cerium oxide (Mohs hardness of 6) were selected as abrasives. Polishing was performed using the following combinations:

- Cast iron–alumina,
- Cast iron–cerium oxide,

- Polyurethane–alumina,
- Polyurethane–cerium oxide.

The results are shown in Fig. 4.17.

The combination of cast iron and alumina showed conchoidal cracks. The combination of cast iron and cerium oxide showed numerous fine cracks and a few scratches. The combination of polyurethane and alumina showed numerous scratches and an orange-peel effect. Only the combination of polyurethane and cerium oxide produced a mirror surface. This indicates that a mirror surface is obtained only when a viscoelastic polisher and soft abrasive particles are used. In such a case, the abrasive particles are embedded in the polisher and scratch the surface of the glass. The polisher acts to transmit the pressure of the load to the abrasive particles. Accordingly, the polisher must have a viscosity that is low enough to allow embedding of the abrasive particles, but must at the same time have a rigidity that is high enough to allow the transmission of the pressure of the load to the abrasive particles. Another necessary function of the polisher is that it must allow viscous flow during polishing so that a process of fitting of the polisher can occur. For such reasons, the polisher must be a viscoelastic body. However, such a function can also be performed by an elastic body such as felt. As will be described later, however, it is difficult to obtain optical polished surface precision with an elastic polisher.

4.1.3.2. Viscoelastic Properties of Polishers

Changes in the creep compliance of representative polishers over time are shown in Figs. 4.18, 4.19, and 4.20. Figure 4.18 shows the creep characteristics of a soft pitch called K 4, and Fig. 4.19 shows the creep characteristics of polyurethane pads; both polishers are viscoelastic bodies. Figure 4.20 shows the creep characteristics of felt, which is an elastic body. Figure 4.21 shows the creep characteristics of various polishers. Polishers that have a creep compliance of 1 to 10×10^{-9} cm^2/dyne are actually used.

4.1.3.3. Relationship of the Rheological Properties of Polishers to the Polishing Rate, Surface Roughness, and Geometrical Surface Precision

Relationship of Rheological Properties of the Polisher to the Polishing Rate. Figures 4.22 and 4.23 show the relationship between polishing

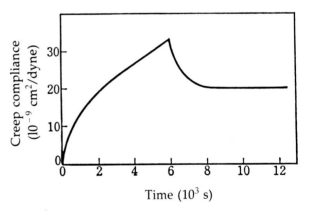

Figure 4.18.   Creep characteristics of K 4 pitch (25°C).

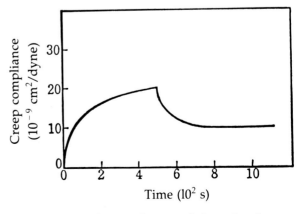

Figure 4.19.   Creep characteristics of polyurethane foam (25°C).

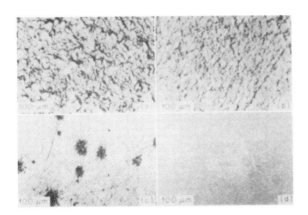

Figure 4.17.   Glass surfaces polished with various polishers and polishing agents: (a) cast iron-alumina, (b) cast iron-cerium oxide, (c) polyurethane-alumina, and (d) polyurethane-cerium oxide.

rate and creep rate for SF 6 and BK 7. For the soft glass SF 6, a maximum polishing rate occurs at a creep compliance value of 30×10^{-9} cm²/dyne; for the hard glass BK 7, a maximum polishing rate occurs at a creep compliance value of 8×10^{-9} cm²/dyne. The increase in the polishing rate that accompanies an increase in the creep extension rate results from an increase in the degree of fitting between the polisher and the lapped surface of the glass. Furthermore, the decrease in the polishing rate that occurs after a maximum is reached results from the relaxation of the load pressure due to lowered viscosity. It is thus indicated that a hard polisher with a low creep extension rate is required for hard glasses, whereas a soft polisher is appropriate for soft glasses.

Relationship of Rheological Properties of the Polisher to Surface Roughness. Figure 4.24A, B, and C shows electron micrographs of polished SF 6 surfaces and of surfaces obtained by etching these polished surfaces with 0.1 N hydrofluoric acid. The figure shows that surface scratches become conspicuously more numerous as the hardness of the polisher increases (in the order K 7 pitch → rosin-impregnated felt).

Geometrical Surface Precision. Figure 4.25 shows a surface that was polished with soft rubber. Disorder is seen in the Newton interference fringes, and ''edge roll off'' is observed. Figure 4.26 shows a surface that was polished with K 7 pitch. There is no disorder at all in the interference fringes, but edge roll off is observed.

However, if the surface is polished with rosin, which has a low creep compliance, the edge roll off also disappears (Fig. 4.27).

The facts indicate that the polisher must be a viscoelastic body (or an elastic body). The polisher acts to embed the abrasive particles; thus causing the abrasive particles to scratch the glass surface and, at the same time, the polisher is fitted to the form of the glass surface. The harder the polisher, the coarser the surface texture. In addition, the geometrical surface precision improves with increased hardness of the polisher. Furthermore, in regard to the polishing rate, an optimal polisher hardness was found to exist for each of the respective glasses.

The idea that polishing is accomplished as a result of ''abrasive particles embedded in the polisher that scratch the hydrated layer on the surface of the glass'' provides a unified, noncontradictory explanation of the ease with which a given glass is polished, the action of the abrasive grains, and the function of the polisher.

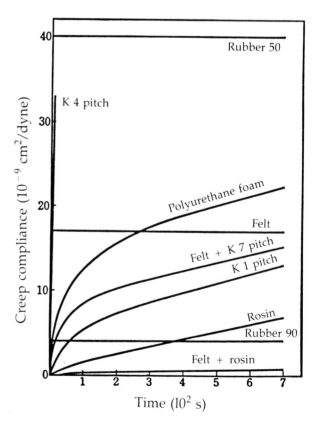

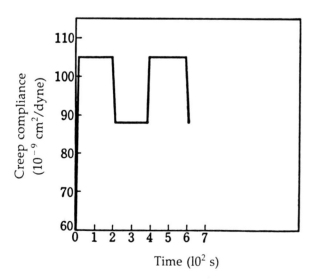

Figure 4.20. Creep characteristics of felt (25°C).

Figure 4.21. Creep characteristics of various polishers (25°C).

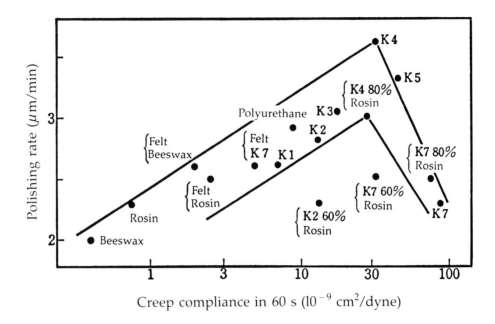

Figure 4.22. Relationship between polishing rate and creep rate in the case of SF 6.

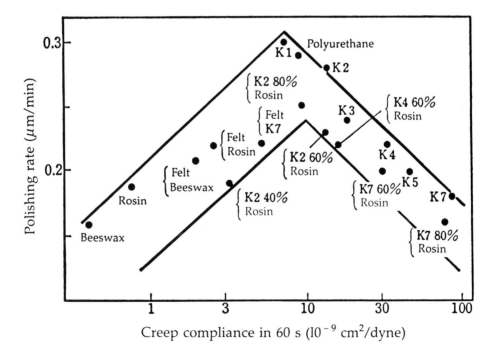

Figure 4.23. Relationship between polishing rate and creep rate in the case of BK 7.

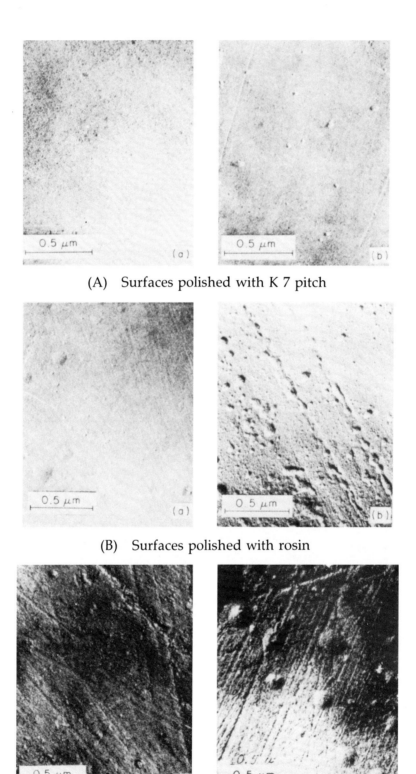

(A)   Surfaces polished with K 7 pitch

(B)   Surfaces polished with rosin

(C)   Surfaces polished with rosin-impregnated felt

Figure 4.24.   Electron micrographs of polished and etched SF 6 surfaces: (a) polished surface and (b) same surface etched with 0.1 N hydrofluoric acid.

Figure 4.25. Surface polished with rubber 40.

Figure 4.26. Surface polished with K 7 pitch.

Figure 4.27. Surface polished with rosin.

4.2. Lapping

Lapping consists of smoothing by means of free abrasive particles, which is in contrast to grinding by means of fixed abrasive particles such as diamond pellets. Lapping hardness expresses the workability of a glass in the lapping process. Ordinarily, however, the hardness of a glass is expressed as an indentation hardness such as Vickers hardness or Knoop hardness. What kind of relationship exists between lapping hardness and indentation hardness? What properties determine the lapping hardness? What is the true nature of the indentation hardness of an optical glass? This section attempts to answer these questions.

In this work, the lapping hardness is the relative value determined from the ratio of the volume decrease of a standard glass to the volume decrease of the sample glass. This value is calculated from the weight loss that occurred after the sample glass (33-mm diameter, 15-mm thickness, and 9-cm^2 lapping surface area) was lapped for a fixed period of time under a fixed load on a cast-iron tool with a lapping mixture of 10 g of Al_2O_3 (20-mm mean grain size) and 20 g of water.

lapping hardness

$$= \frac{\text{volume decrease of standard glass}}{\text{volume decrease of sample glass}} \times 100$$

$$(4.1)$$

Figure 4.28 shows the relationship between the Knoop hardness and lapping hardness of optical glasses. In general, a correspondence exists between the two hardnesses; however, there is a considerable breadth in the proportional zone so that glasses with the same Knoop hardness may show a considerable difference in lapping hardness. In other words, the lapping hardness cannot be immediately predicted from the Knoop hardness.

4.2.1. Indentation Hardness of Optical Glasses

There are two theories regarding the indentation hardness of glasses. One is the densification theory[14-16]; the other is the microscopic plastic flow theory.[17,18] In the case of glasses such as optical glasses that contain large amounts of modifier oxides, however, the impression made by the indenter appears to result from plastic flow. Figure 4.29 is a reflecting interference micrograph of a Knoop indentation in SF 5 (load: 1 kg·15 s). As seen from the figure, the interference fringes (around the periphery of the indentation) run in the opposite direction from the interference fringes inside the indentation, and bumps are formed around the periphery of the indentation. Figure 4.30 shows a cross section of the indentation: the depth is 3.3 μm, the width is 85 μm, and the bumps are 0.3 μm high. The cross-sectional area of the indented pit is equal to the cross-sectional area of the bumps, indicating that the impression is formed by plastic flow. Similar bumps are found in BK and SK glasses, indicating that the indenter impressions in optical glasses are formed by plastic flow.

4.2.2. Lapping Hardness

4.2.2.1. True Nature of Lapping Hardness

Figure 4.31 shows the relationship between lapping hardness and the length of cracks generated from a Vickers indentation. When a 200-g load is applied for 15 seconds, cracks extend from the corners of the indentation as shown in Fig. 4.32; the extension of these cracks reaches an equilibrium approximately 5 seconds after removal of the load. Crack length was measured using a 400-power microscope. Figure 4.31 shows that there is a linear relationship between lapping hardness and crack length, indicating a correspondence between lapping hardness and the extension of cracks generated from a Vickers indentation. In other words, lapping hardness appears to indicate the degree of resistance to cracking. It is thought that in the lapping process, the lapping abrasive particles form indentations (instead of a Vickers indenter), and minute cracks are generated from these indentations. This agrees with the theory of Imanaka,[19] in which the lapping of a glass is viewed as an aggregation of cracks. Furthermore, the length of the indentation made by a hardness tester may be thought of as corresponding to the size of the indentations made by the lapping abrasive particles.

Lapping hardness is not really a "hardness," but rather a property that depends primarily on the strength of the glass. Lapping hardness is determined by the size of the indentations made by the lapping abrasive particles, and by the extension of cracks from these indentations. Accord-ingly, lapping hardness may be characterized as a property that depends on the indentation hardness of the glass and the strength of the glass.

4.2.2.2. Lapping Hardness and Mechanical Strength

There are two theories regarding the mechanical strength of glasses: the Orowan–Griffith theory,[20,21] in which the glass is viewed as a perfectly elastic body, and the Marsh theory,[22] in which the strength is determined by the threshold value of plastic flow; however, there is a correspondence between the two theories. In this work, the relationship between the strength of the glass and the lapping hardness was determined by calculating the threshold value from the indentation hardness and Young's modulus of the glass using Hill's theory,[23] and taking this as the strength of the glass.

As shown in Fig. 4.33, the volume decrease caused by lapping decreases with an increase in the strength of the glass, though there is quite a bit of variation. If oil is used instead of water, this correspondence becomes closer, and if a dry nitrogen atmosphere is substituted, a clear inversely proportional relationship appears (Fig. 4.34). At the same time, the volume decrease caused by lapping drops by a fair amount compared with the case where water was used. This corresponds to the stress-corrosion theory of Charles and Hillig,[24] and is interpreted as follows: even at the same mechanical strength, the fracture strength differs if the chemical durability differs so that the variation shown in Fig. 4.33 is generated. If, however, the influence of moisture is completely eliminated, then an underlying inversely proportional relationship appears. The fair amount of spread

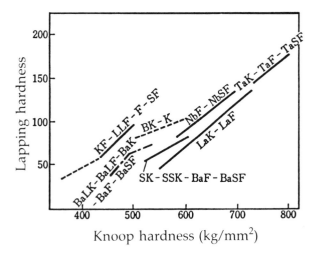

Figure 4.28. Knoop hardness and lapping hardness of optical glasses.

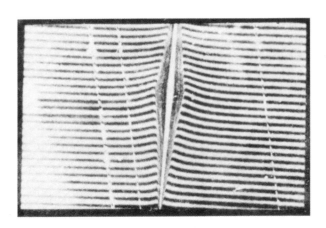

Figure 4.29. Reflecting interference micrograph of Knoop indentation.

that occurs results from differences in the size of the indentations. Furthermore, Fig. 4.35 shows the relationship between the length of the Knoop indentation and the volume decrease caused by lapping. A proportional relationship exists between the two values; the fact that two glasses with the same indentation length will show different lapping volume decreases results from differences in the strengths of the respective glasses. The fact that SF glasses, which contain large amounts of lead, show an inversely proportional relationship in this case is interpreted as follows: because these glasses contain large amounts of polar Pb^{2+}, they have great plasticity so that the concentration of stress at the leading ends of cracks is relaxed.[25]

Lapping strength is a measure of the ease with which a glass is fractured by the lapping abrasive particles and is a property that depends on the chemical durability, hardness, and mechanical strength of the glass involved.

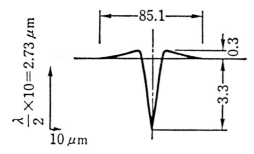

Figure 4.30. Cross-sectional diagram of Knoop indentation.

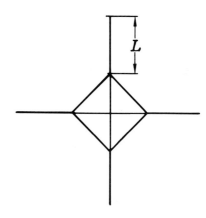

Figure 4.32. Vickers indentation and cracks generated from this indentation.

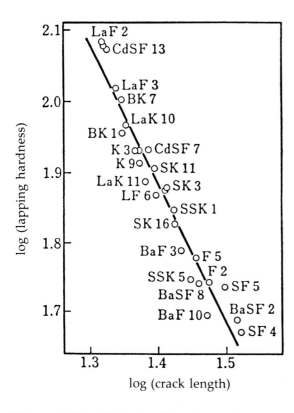

Figure 4.31. Relationship between lapping hardness and length of cracks generated from a Vickers indentation.

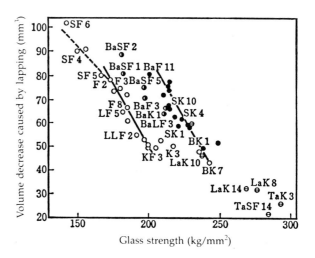

Figure 4.33. Relationship between glass strength and lapping volume decrease in the case of lapping using water.

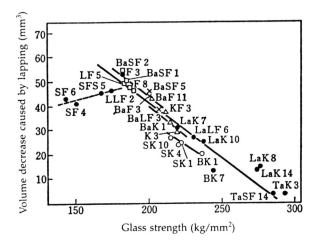

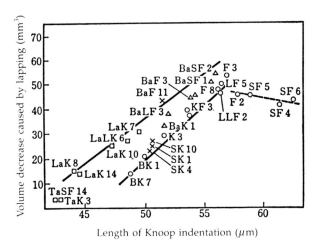

Figure 4.34. Relationship between glass strength and lapping volume decrease in the case of lapping using oil in a nitrogen atmosphere. Indentation size: ○ 50.0 to 50.6, △ 51.7 to 52.1, × 53.9 to 54.2, □ 56.0 to 56.5.

Figure 4.35. Relationship between length of Knoop indentation and volume decrease caused by lapping.

4.3. Diamond Grinding

The roughing process for lenses and prisms generally consists of rough grinding using a diamond tool. In recent years, fine grinding by means of diamond pellets is also beginning to be used in the smoothing process instead of using lapping by means of free abrasive particles. Diamond pellet grinding differs from lapping with free abrasive particles in the following respects: (1) fixed abrasive particles are used, and (2) a metal bond is used.

4.3.1. Mechanism of Grinding

To elucidate the mechanism of diamond grinding, microscopic surface observations were made during the grinding process. In addition, changes in the amount of grinding over time, changes in surface roughness, and changes in the amount of wear of the diamond pellets were also investigated. The mechanism of diamond grinding cannot be explained without clarifying the effect of the metal bond. Accordingly, the effect of the metal bond was also investigated.

For the above investigations, eight types of diamond pellets were prepared by mixing diamond abrasive particles [grain size #1500 (5 to 12 μm)] with Cu–Sn–Fe type metal bonds. The mechanical properties of the bonds are shown in

Fig. 4.36 and Table 4.3. The figure shows the relationship between the Rockwell hardness and the bending strength of the bonds, and the table shows the degree of abrasive wear of the bonds. From the figure and the table, we see that the degree of wear of the bond depends more on the bending strength of the bond than on the hardness, and that the bond wears through mechanical fracture. Metal bonds in which the degree of wear decreased (in the order No. 2 through No. 8) were prepared to ascertain the relationship with the mechanism of diamond grinding. The glasses selected for use in these tests were BK 7, F 2, SF 6, SK 16, BaSF 8, LaK 12, and TaF 1. In this way, glasses with different Knoop hardnesses (350 to 750 kg/mm^2) and degrees of abrasive wear (60 to 230) were included. The pellets thus manufactured were pasted to cast-iron tools with epoxy resin; the grinding experiments were performed using an Oskar type polisher with water fed at 2.5 liters/minute as a grinding liquid.

4.3.1.1. Changes in the Cumulative Amount of Grinding over Time

As shown in Fig. 4.37, three types of relationships are observed between the amount of grinding and the length of grinding time according to the type of glass and metal bond involved (see Figs. 4.38 through 4.42).

● Type A: the amount of grinding varies linearly with time; when the glass surface is observed with a microscope, the surface is found to be covered with conchoidal fractures (Fig. 4.43A). This indicates that the glass is removed by fracture in this case.

● Type B: the relationship between the amount of grinding and time is a curved line. In the initial stages of grinding, the glass surface is covered with conchoidal fractures; as grinding progresses, however, scratches increase (Fig. 4.43B).

● Type C: the relationship between the amount of grinding and time is linear except for the initial stage; the surface is covered with scratches (Fig. 4.43C). The mechanism of removal is plastic scratching.

In conclusion, there are basically two types of diamond grinding mechanisms:

● Fracturing: this is exactly the same as the mechanism of lapping.

● Plastic scratching: this is not seen in lapping; it is a phenomenon that appears when the metal bond is strong and the hardness of the glass surface is soft. In this type, the diamond pellets consist of fixed abrasive particles. As the metal bond becomes stronger, almost all the glasses

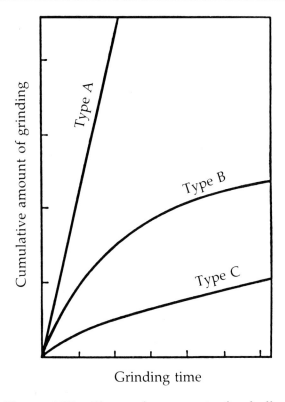

Figure 4.37. Change in amount of grinding over time.

Table 4.3. Degree of abrasive wear of bonds.

Bond No.	1	2	3	4	5	6	7	8
Degree of wear	5.5	10.3	7.0	4.8	4.7	4.4	4.0	4.0

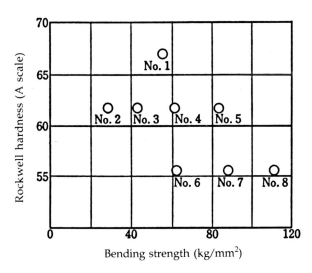

Figure 4.36. Rockwell hardness and bending strength of bonds.

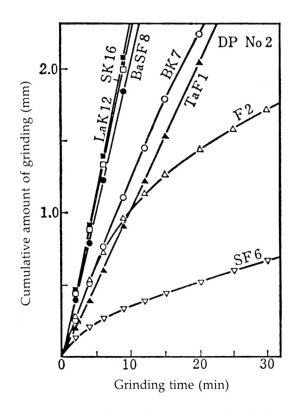

Figure 4.38. Change in amount of grinding over time.

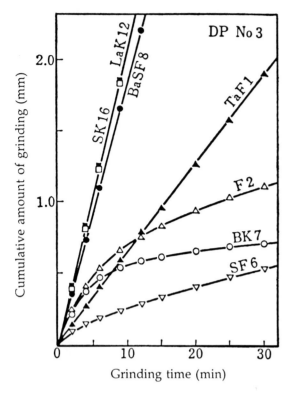

Figure 4.39.   Change in amount of grinding over time.

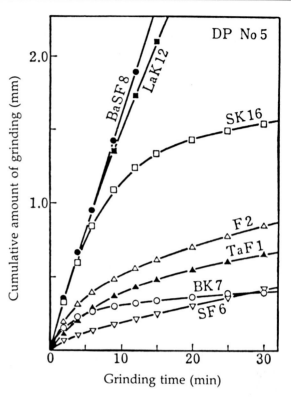

Figure 4.41.   Change in amount of grinding over time.

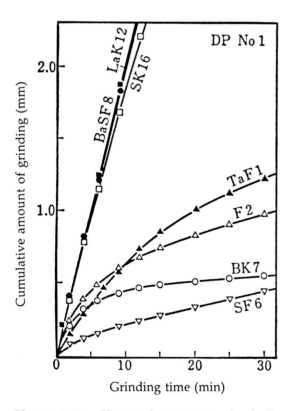

Figure 4.40.   Change in amount of grinding over time.

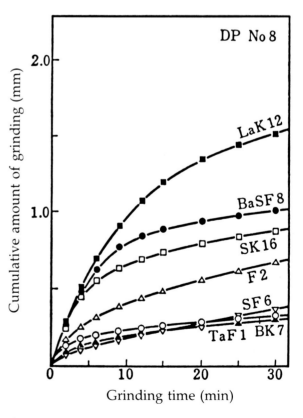

Figure 4.42.   Change in amount of grinding over time.

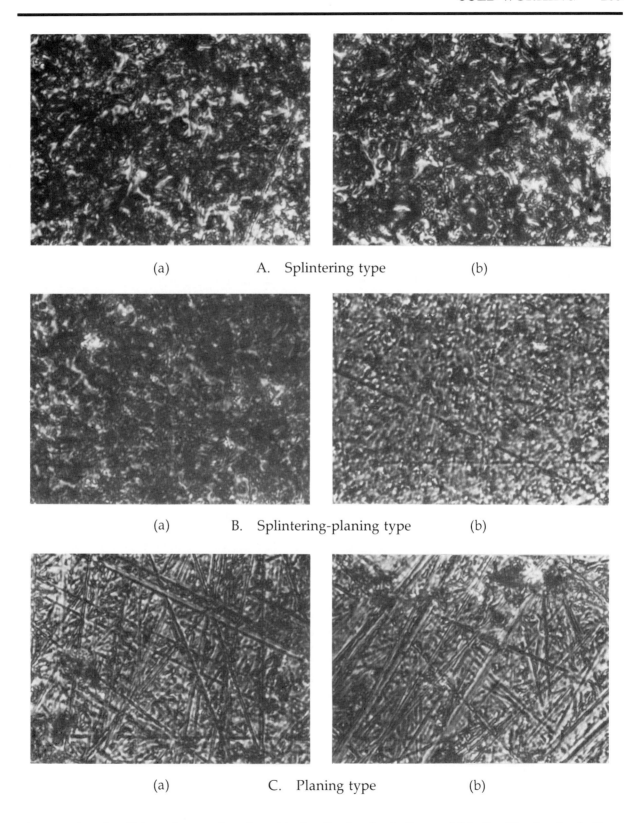

(a)   A. Splintering type   (b)

(a)   B. Splintering-planing type   (b)

(a)   C. Planing type   (b)

Figure 4.43. Condition of ground surface: (a) after 2 min of grinding and (b) after 12 min of grinding.

show type B and type C changes in the amount of grinding over time (see Fig. 4.41).

4.3.1.2. Changes in Surface Roughness over Time

Figures 4.44 through 4.48 show changes in surface roughness over time. When the metal bond is weak, the change in surface roughness over time also shows three types of variation (see Fig. 4.49).

• Type a: the roughness does not change. This is interpreted as indicating the occurrence of glass fracturing regardless of time elapsed.

• Type b: the roughness decreases with time. The reason for this is that the removal mechanism shifts from fracturing to plastic scratching.

• Type c: the roughness decreases abruptly immediately after grinding is begun, and then shows almost no change. This occurs because the removal mechanism is plastic scratching; dressing of the metal bond by glass fractures does not take place except in the period immediately following the initiation of grinding.

As the metal bond becomes stronger, scratching occurs in all the glasses so that dressing of the metal bond no longer takes place. As a result, the change in surface roughness shifts to type b or type c.

4.3.1.3. Changes in the Amount of Diamond Pellet Wear over Time

Figures 4.50 through 4.54 show the relationship between the amount of diamond pellet wear and time. As shown in Fig. 4.55, the following two types of relationships are observed:

• Type α: the diamond pellets wear linearly with time. This indicates that the pellets are constantly being dressed by glass grinding splinters.

• Type β: the diamond pellets wear only during the period immediately following the initiation of grinding; after this period, almost no pellet wear is observed. This indicates that almost no dressing occurs because the removal mechanism is plastic scratching.

As the metal bond becomes stronger, glass is removed almost exclusively by scratching. As a result, dressing does not occur, and the diamond pellets do not wear.

From the above discussion, the following three types of diamond-pellet grinding mechanisms can be distinguished:

• Removal by fracturing: because glass is removed by splintering, both the amount of glass ground and the amount of wear of the diamond

pellets vary linearly with time; the surface roughness of the glass stays constant.

• Removal by fracturing and scratching: the removal of glass is accomplished by both fracturing and scratching. Fracturing occurs in the initial stages of grinding; as grinding progresses, removal by scratching increases. Because fracturing occurs in the initial stages in which a rough glass surface is ground, diamond-pellet wear is greatest in the initial stages. In the subsequent stages of the process, dressing does not occur; accordingly, the grinding performance drops, the amount of glass ground decreases, and the surface roughness also decreases.

• Removal by scratching: in this case, plastic scratching occurs from the initial stages of grinding. The amount of grinding is small; the surface roughness is low and becomes fixed after a given period of time. Furthermore, since dressing does not occur except in the period immediately following the initiation of grinding, the diamond pellets do not wear.

4.3.1.4. Effect of the Metal Bond on the Grinding Rate

For diamond pellets that use a bond with a low degree of abrasive wear and strong bonding, the grinding rate required to produce scratches in the glass is low. Conversely, for diamond pellets that use a bond with a high degree of abrasive wear and weak bonding, the grinding rate required to produce splintering in the glass is high.

In diamond grinding, both the holding force of the diamond abrasive particles (which is proportional to the strength of the bond) and the ease with which dressing occurs (which is inversely proportional to the strength of the bond) would appear to have an effect on the grinding rate. In the case of ordinary diamond pellets, however, the holding force is sufficient so that the grinding rate depends on the ease with which the diamond pellets are dressed.

As Fig. 4.56 shows, the rate of lapping (degree of abrasion) by free abrasive particles and the rate of diamond grinding (initial grinding rate) do not necessarily show a correspondence because the glass removal mechanisms involved are not necessarily the same. In the case of diamond grinding, the relative initial grinding rates for various types of glass show almost no change even if the mechanical properties of the bond differ. This occurs because the mechanism of removal in initial-stage grinding is determined by the type of glass and does not depend on the strength of the metal bond.

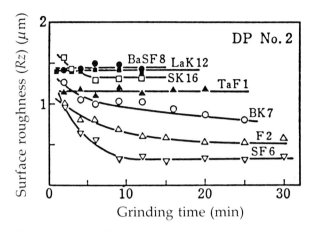

Figure 4.44.   Change in surface roughness over time in the case of DP No. 2.

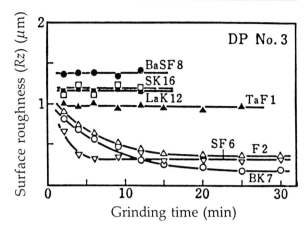

Figure 4.45.   Change in surface roughness over time in the case of DP No. 3.

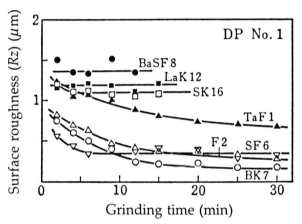

Figure 4.46.   Change in surface roughness over time in the case of DP No. 1.

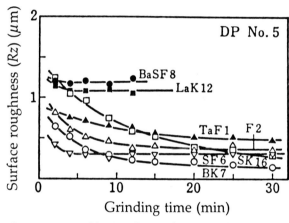

Figure 4.47.   Change in surface roughness over time in the case of DP No. 5.

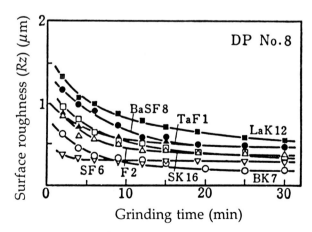

Figure 4.48.   Change in surface roughness over time in the case of DP No. 8.

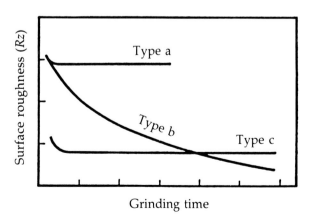

Figure 4.49.   Change in surface roughness over time.

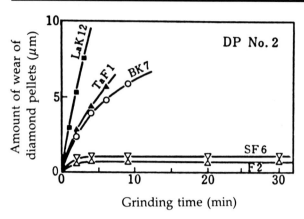

Figure 4.50.   Change in the amount of wear of diamond pellets over time in the case of DP No. 2.

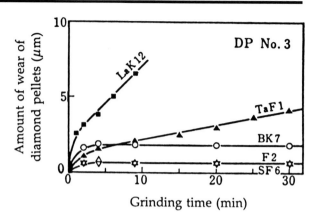

Figure 4.51.   Change in the amount of wear of diamond pellets over time in the case of DP No. 3.

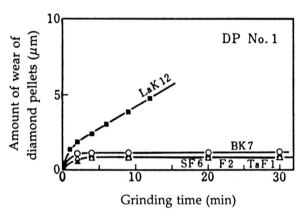

Figure 4.52.   Change in the amount of wear of diamond pellets over time in the case of DP No. 1.

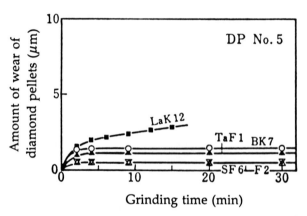

Figure 4.53.   Change in the amount of wear of diamond pellets over time in the case of DP No. 5.

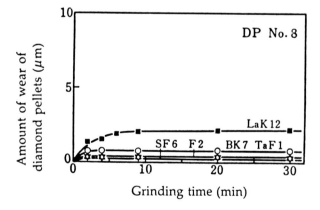

Figure 4.54.   Change in the amount of wear of diamond pellets over time in the case of DP No. 8.

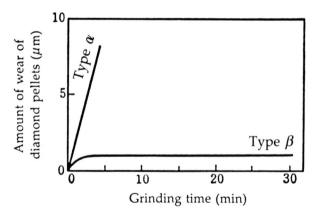

Figure 4.55.   Change in the amount of wear of diamond pellets over time.

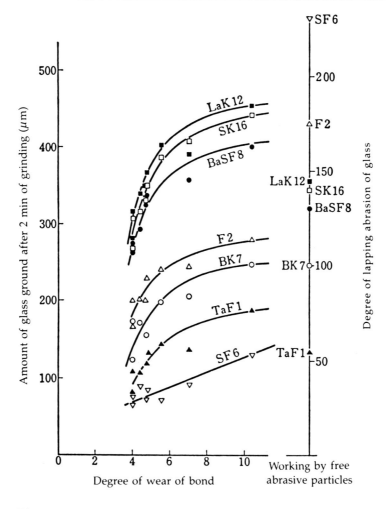

Figure 4.56. Relationship of bond strength to diamond grinding rate and glass lapping rate.

4.3.1.5. Relationship between Diamond Grinding Rate and Lapping Rate

Figure 4.57 shows the relationship between the grinding rate obtained using diamond pellets with a mean grain size of 80 μm, and the lapping rate obtained using free abrasive particles of SiC with a mean grain size of 34 μm. Figure 4.58 also shows the relationship between the grinding rate obtained using diamond pellets with a mean grain size of 20 μm and the lapping rate obtained using SiC with the same mean grain size as the diamond pellets.[26] A proportional relationship exists between the two rates when the abrasive grain size is large because both grinding and lapping proceed by splintering of the glass in this case. As the abrasive grain size becomes smaller, however, no such proportional relationship is shown for heavy flint glasses: the diamond grinding rate is lower than the lapping rate because the latter rate de-

pends on a splintering mechanism, whereas the former rate depends on a scratching mechanism. Thus, because the lapping rate and the diamond grinding rate do not necessarily depend on the same removal mechanism, they do not always show a correspondence.

4.3.1.6. Summary

Diamond grinding mechanisms include both removal by splintering and removal by scratching. The differences between these modes of removal depend on the properties of the glass itself, the size of the diamond abrasive particles, the mechanical properties of the metal bond of the diamond pellets, and the mechanical conditions of grinding such as load, etc. The grinding rate and surface roughness are affected by the mode of glass removal. When the removal mechanism is splintering, the diamond grinding rate is proportional to the lapping rate. When removal by scratching occurs, however, no correspondence exists between the two rates. In addition to diamond-abrasive-particle wear, the presence or absence of a dressing effect plays a primary role in determining the change in the grinding rate over time. When the removal mechanism is splintering, there is a dressing effect by ground glass powder; the magnitude of the effect would appear to depend on the size of the ground powder. When the removal mechanism is scratching, no dressing effect occurs.

4.3.1.7. Grinding Liquid Used in Diamond Grinding

In general, water is not used as a grinding liquid in diamond grinding. It is more common to use a grinding liquid containing ethylene glycol, etc.; Fig. 4.59 shows that the grinding rate is higher when this type of grinding liquid is used. The reason for this follows: when water is used, scratching occurs in addition to fracturing (as seen from the change in the amount of grinding over time). In contrast, when a grinding liquid such as that mentioned above is used, the amount of

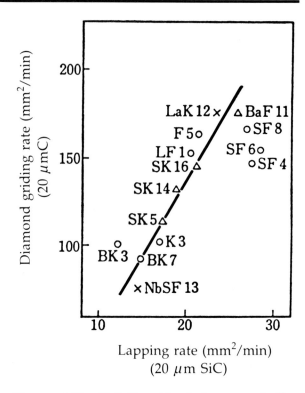

Figure 4.58. Relationship between grinding rate and lapping rate in the case of small abrasive particles.

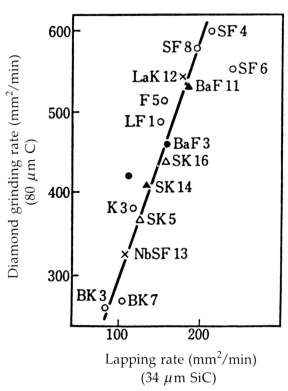

Figure 4.57. Relationship between grinding rate and lapping rate in the case of large abrasive particles.

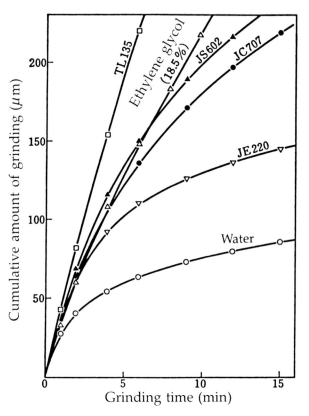

Figure 4.59. Effect of grinding liquid on diamond grinding.

grinding increases linearly with time, indicating that only fracturing occurs. The reason for this is inferred to be as follows: when water is used, a hydrated layer is formed on the surface that leads to scratching. On the other hand, when a grinding liquid such as that described above is used, there tends to be no formation of a hydrated layer so that only fracturing occurs.

4.4. Dimming and Staining in the Polishing Process

One problem that occurs in the optical-glass polishing process is the problem of dimming and staining. The nature and causes of dimming and staining, and the methods used in testing for dimming and staining are discussed in Chapter 2. Although the basic phenomena that underlie dimming and staining are fairly clear, the production of unsatisfactory products resulting from dimming and staining is still a problem in lens polishing. If we assume that dimming and staining are essentially unavoidable in ordinary polishing processes, it appears that we cannot completely solve the problem unless we resort to an improvement of the chemical durability of the glass materials themselves. Therefore, in this study, we tried to ascertain exactly what parts of the polishing process tend to generate dimming and staining, and at the same time, investigated the problem of whether or not dimming and staining are avoidable. In the experiments, we used SK 16, which has an extremely poor chemical durability and is said to be susceptible to dimming and staining.

Figure 4.60 shows a standard lens polishing process. We investigated the generation of dimming and staining in the various processes according to this process diagram. To facilitate the detection of dimming and staining, half of the surface of each glass sample was coated with a protective film; after the completion of a specified process, this protective film was removed using thinner. Staining was detected by means of the staining detector shown in Fig. 4.61. This device uses the difference in the Brewster angle between the film and the glass[27] (see Fig. 4.62). When no staining

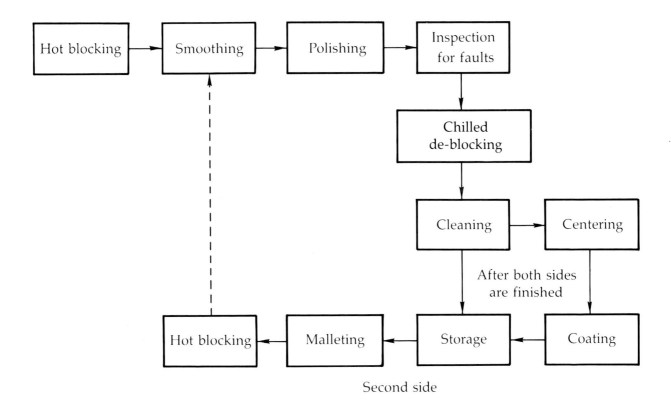

Figure 4.60. Lens polishing process.

exists, no visible boundary shows between the portion of the glass covered by a protective film and the uncovered portion of the glass. Accordingly, the visual field appears uniformly dark. In contrast, when staining is generated, the area where staining occurred looks lighter, whereas the area where no staining was generated (i.e., the area covered by a protective film) appears darker. The degree of staining is indicated by the symbols shown in Table 4.4. Dimming was detected by means of an electron microscope.

4.4.1. Polishing Process and Wiping Process Following the Completion of Polishing

The question for this study was whether or not dimming and staining were generated during polishing; as shown in Fig. 4.63, neither dimming nor staining occurs. After the completion of polishing, the polished lenses are wiped with a damp cloth to remove any polishing solution. Next the lenses are sent through the fault inspection process because there is a possibility that some polishing liquid may remain on the lens surface, or that water remaining in the mallet area may become trapped between the glass and the protective film when the protective film is applied to the lens. Figure 4.64 shows a case where polishing liquid remaining on the polished surface was naturally dried for approximately 15 minutes. Both staining and dimming occur on the surface. Figure 4.65 shows the results obtained when oil droplets were applied to the polished surface and naturally dried along with the remaining moisture. Dimming occurs in annular patterns around the oil droplets on the surface. From these results, we can say that dimming and staining occur in areas where water, polishing solution, contamination, etc., remain on the surface after polishing.

4.4.2. Protective Film Coating Process

Lenses that have passed through the fault inspection process are coated with a protective film and are sent to the de-blocking process. When polished surfaces from which moisture had been eliminated were coated with acrylic resin protective films (strong) and pitch protective films and were allowed to stand for 7 days in the polishing room, no dimming or staining occurred. Accordingly, we may say that no dimming or staining will occur for at least a week if the water-free surface is coated with a protective film.

4.4.3. De-Blocking Process

Lenses that have passed through the fault inspection process and that have been coated with a protective film are usually cooled to approximately −30°C by means of a refrigerator and then removed from the block holder and mallets. The alcohol in the refrigerator commonly contains 10 to 40% water. In addition, the protective film on the surface of each lens often cracks or peels during cooling. In such cases, the bare lens surface is exposed to water. Furthermore, water vapor condenses on the lens surface for a considerable

Table 4.4.   Indication of degree of staining.

Symbol	Meaning
○	No staining
◑	Weak staining generated
●	Strong staining generated (visible to the naked eye)

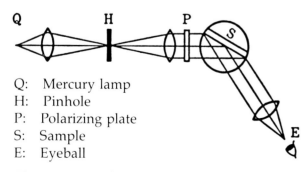

Q: Mercury lamp
H: Pinhole
P: Polarizing plate
S: Sample
E: Eyeball

Figure 4.61.   Principle of the staining detector.

(a) No staining present   (b) Staining present

Figure 4.62.   Example of staining observation.

period of time after the lens has left the refrigerator; this condensation continues until the lens surface reaches room temperature. The lens is allowed to stand in this condition until the next process begins. Accordingly, it appears that dimming and staining occur fairly easily in this process.

When bare lenses (with no protective film) were passed through a refrigerator and were then allowed to stand for 30 minutes until the lenses reached room temperature, both staining and dimming occurred on the lens surfaces. Figure 4.66 shows the results obtained when a lens was coated with a pitch protective film, and then was allowed to stand for 30 minutes after a refrigeration treatment. Figure 4.67 shows the re-

sults obtained when a lens surface subjected to the same treatment was allowed to stand for 16 hours with water droplets adhering to the surface. Although the pitch protective film showed no cracks or spots of peeling, a slight amount of dimming was observed. Allowing the glass to stand caused the dimming to grow even further, and also caused staining to appear. Neither staining nor dimming occurred on a lens surface that was coated with an acrylic-resin protective film, passed through the refrigerator, and allowed to stand for 16 hours with water droplets adhering to the surface (Fig. 4.68). Judging from these results, it appears that dimming and staining can be prevented in this process if the glass surface is coated

Figure 4.63. Surface immediately after polishing.

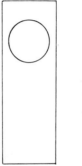 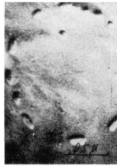

Figure 4.65. Surface on which oil droplets and polishing liquid were naturally dried.

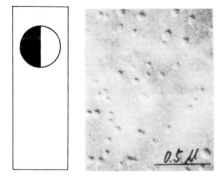

Figure 4.64. Surface on which polishing liquid was naturally dried for 15 min.

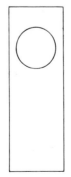 

Figure 4.66. Surface coated with a pitch protective film, passed through a refrigerator, and allowed to stand for 30 min.

with a protective film that has strong water-resistant properties and will not peel when rapidly cooled.

4.4.4. Cleaning Process

Lenses that have been removed from the block holder and mallets by de-blocking under refrigeration are placed in an ultrasonic cleaner, where adhering pitch and polishing agent, etc., are cleaned away. Cleaning is performed in the sequence shown in Fig. 4.69. The lens immersion time in each cleaning tank is 50 seconds. In this process (unlike other processes), the lens is constantly in contact with large amounts of liquids; therefore, there is a very real possibility of staining or latent scratching.

Figure 4.70 shows a polished surface before it was placed in the cleaning apparatus. This sample surface happened to have slight dimming present. Figure 4.71 shows the results obtained after trichloroethylene cleaning in tank No. 1 and tank No. 2. As before, a slight amount of dimming is present on the surface. Figure 4.72 shows the results obtained after cleaning with water in tank No. 3. The dimming that was present before cleaning has dissolved and disappeared, and no staining has been generated. Figure 4.73 shows the results obtained after cleaning with a neutral detergent (pH 11) in tank No. 4. Here, scratches are observed on the surface. The results obtained after cleaning with alcohol in tank No. 5 and tank No. 6 show almost no change from the results obtained after cleaning in tank No. 4. It appears from these results that neither staining nor dimming is

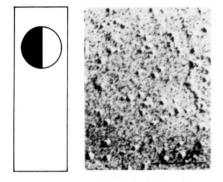

Figure 4.67. Surface coated with a pitch protective film, passed through a refrigerator, and allowed to stand for 16 hr.

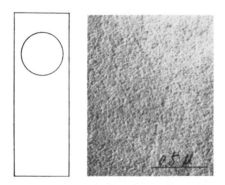

Figure 4.68. Surface coated with an acrylic-resin protective film, passed through a refrigerator, and allowed to stand for 16 hr.

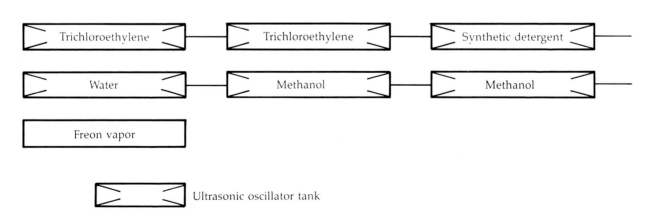

Figure 4.69. Lens cleaning process.

generated in the trichloroethylene tanks or alcohol tanks, and that reactions between the glass and organic solvents such as trichloroethylene, alcohol, etc., can be ignored.

Weak staining was generated in tank No. 4; however, this was generated as a result of erosion by the water in tank No. 3 and the neutral detergent in tank No. 4, and was probably generated primarily as a result of selective leaching by the water in tank No. 3. Scratches (latent scratches) were observed in the neutral detergent tank; however, these resulted from scratches that were formed during the polishing process and enlarged by erosion. The action of ultrasonic waves appears to promote both the generation of staining by the chemical action of water and the spread of scratches developed by the neutral detergent. In any case, leaving differences in degree aside, note that staining is inevitably generated in this process.

4.4.5. Malleting and Hot Blocking

Malleting precedes the working of the second side of the lens and is a process in which melted pitch is poured onto the first side of the lens (which has already been coated with a protective film after polishing) so that a mallet can be fixed to the lens. Hot blocking is a process in which a tool heated to approximately 200°C is pressed against the malleted lenses so that the lenses are fixed to the tool. After hot blocking is completed, the lenses fixed to the tool are immersed in water for the purpose of cooling. It is conceivable, however, that the protective film and/or mallet might not adhere perfectly to the lens, or that they might peel off during the treatment.

When the hot blocking process is completed with the lens surface completely covered by the protective film and mallet, neither staining nor dimming occurs. Figure 4.74 shows results for a

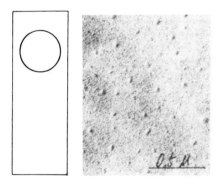

Figure 4.70. Surface before cleaning.

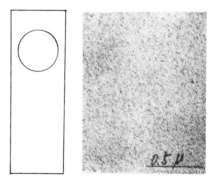

Figure 4.72. Surface after cleaning with water.

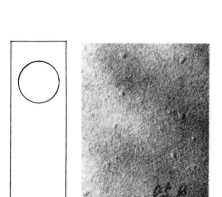

Figure 4.71. Surface after cleaning with trichloroethylene.

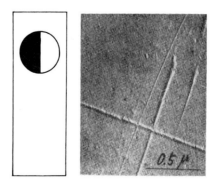

Figure 4.73. Surface after cleaning with a neutral detergent.

portion of a lens surface where the hot blocking process was completed with only a strong protective film adhering to the surface; the mallet having peeled away. Dimming is generated on the surface, though to an extremely slight degree. Figure 4.75 shows results obtained in a case where the hot-blocking process was completed with neither a protective film nor a mallet adhering to the lens surface; both staining and dimming occur on the surface. Judging from these results, it appears that neither dimming nor staining will be generated if the hot-blocking process is completed with both the protective film and mallet adhering perfectly to the entire surface of the lens.

4.4.6. Centering Process

Lenses that have been cleaned on both sides are sent to the centering process. Here, lamp oil (kerosene) is commonly used as a grinding oil. In our experiments, each lens was immersed in kerosene for 24 hours, but neither staining nor dimming appeared. Figure 4.76 shows the results obtained when water droplets were deliberately applied to the lens surface before the lens was immersed in kerosene for 24 hours. In this case, staining appears on the surface. Judging from these results, it is essential to avoid the adhesion of water droplets to the lens surface before the lens comes into contact with the grinding oil in the centering process.

4.4.7. Standing in the Shop

After both sides are polished and cleaned, the lenses are wiped and sent to the centering process. However, between the completion of this work and the coating of the lenses, the lens surfaces may be left standing exposed in the shop for a long period of time (about two days).

Figure 4.77 shows the results obtained when a lens was allowed to stand for 48 hours in a desiccator at a temperature of 25°C and a humidity of 50%. Here, no dimming or staining appears on the surface. Figure 4.78 shows the results obtained when a lens was allowed to stand for 24 hours in a desiccator at a temperature of 25°C and a humidity of 90%. In this case, only dimming appears on the surface. Figure 4.79 shows the results obtained when a lens was allowed to stand for 60 minutes at a temperature of 34°C, a humidity of 100%, and with water droplets condensing on the lens surface. In this case, both staining and dimming appear on the surface. These results indicate that the tendency for dimming to occur when a lens is left standing in an exposed condition and the underlying cause of this tendency can be explained in terms of high indoor humidity and the condensation of moisture in the air on the lens surface. Consequently, the maintenance of a constant temperature and a low humidity in the shop would appear to be a prerequisite for the prevention of dimming and staining.

4.4.8. Summary

For various processes, we investigated the occurrence of dimming and staining in the lens polishing process for SK 16 glass. We found that the dimming and staining phenomena tend to occur in the following processes:

1. During the wiping of the lens surface, following polishing.
2. During de-blocking under refrigeration, as a result of peeling of the protective film.
3. During cleaning, as a result of reaction with water.
4. During hot blocking, as a result of faulty adhesion of the protective film and mallet.
5. During standing in the shop before and after centering, and during standing in the shop before coating.

The following may be cited as examples of countermeasures that can be used to prevent dimming and staining in the above processes:

1. Wiping after polishing: a method must be devised to immediately and completely wipe away all moisture.
2. Solving the problem of protective film: a material must be used that has strong water-resistant properties and will not crack or peel away from the lens surface at low temperatures, e.g., −30°C. One essential condition is that no moisture be trapped under this protective film when the film is applied.
3. Cleaning: the lens is extremely susceptible to staining; though the degree may vary, the generation of some staining is inevitable. Accordingly, the only remedy for this problem is to improve the chemical durability of the glass.
4. Hot blocking: this problem can be solved if the process is performed so that the protective film and mallet adhere perfectly to the surface of the lens.
5. Dimming: this is most likely to occur while the glass is standing in the shop. Because it is necessary to prevent the condensation of moisture; a constant temperature must be maintained

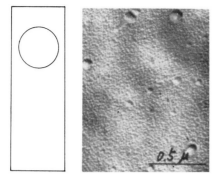

Figure 4.74. Surface where the hot blocking process was completed with only a protective film adhering to the surface.

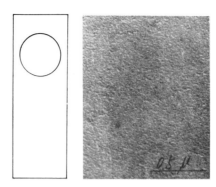

Figure 4.77. Surface that was allowed to stand for 48 hr at a temperature of 25°C and a humidity of 50%.

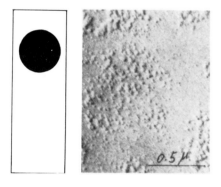

Figure 4.75. Surface where the hot blocking process was completed with neither a mallet nor a protective film adhering to the surface.

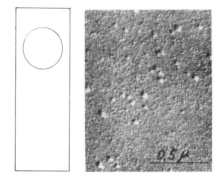

Figure 4.78. Surface that was allowed to stand for 24 hr at a temperature of 25°C and a humidity of 90%.

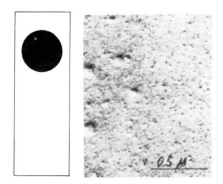

Figure 4.76. Surface that was immersed in kerosene for 24 hr after water droplets were applied.

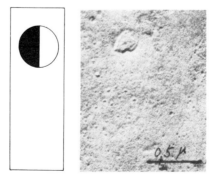

Figure 4.79. Surface that was allowed to stand for 60 min at a temperature of 34°C and a humidity of 100% (condensation occurring).

and a high relative humidity must be avoided in the shop.

If the principle of avoiding contact between the lens surface and water (and avoiding the condensation of water vapor on the lens surface) is observed in order to prevent dimming and staining as described above, it appears that dimming and staining can be prevented even in glasses that have a fairly poor chemical durability. However, contact with water cannot be avoided in the cleaning process; accordingly, it is the responsibility of the glass maker to make a glass that can withstand at least 50 seconds of contact with water in ultrasonic cleaning.

4.5. Cleaning

In the glass polishing process, cleaning is performed a number of times. Furthermore, this cleaning is closely connected with the subsequent process of coating. Here, we discuss the following problems: (1) What contamination is generated on the polished surfaces of optical glasses? (2) What are the causes of this contamination? (3) How can this contamination be removed? We also discuss surface deterioration (staining and latent scratching) accompanying cleaning, and the relationship between cleaning and the adhesive strength of coatings deposited by vacuum evaporation.

4.5.1. Contamination Generated on the Polished Surfaces of Optical Glasses and Causes of this Contamination

The general process by which optical glasses are manufactured is described in a previous section; for ease of understanding, however, this process is shown again in Fig. 4.80. The types of contamination generated in the various processes are as follows:

1. After de-blocking (cleaning I-II): contamination found on the polished surface after

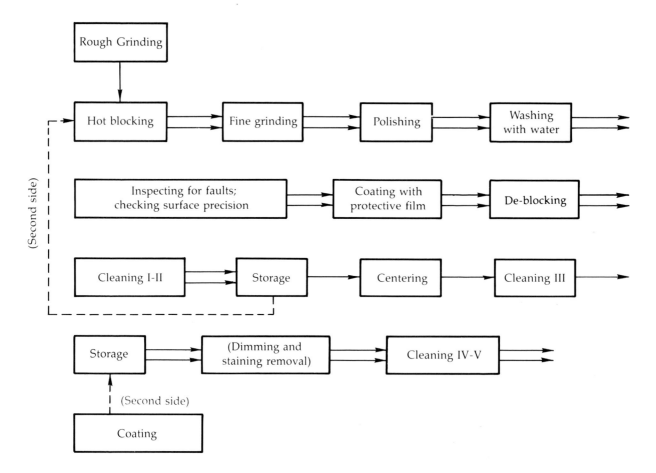

Figure 4.80. Optical lens manufacturing process.

de-blocking consists of pitch used for fixing the glass, protective film material, fingerprints, polishing agent (CeO_2, ZrO_2, etc.), dimming generated on the polished surface, dust floating in the workshop, etc.

2.   After centering (cleaning III): grinding oil, fingerprints, ground glass powder, dimming, dust, etc., are present as contamination on the polished surface after centering.

3.   After removal of dimming and staining (cleaning IV–V): a dimming and stain-removal process is performed only for glasses that are susceptible to dimming and staining. Soft polishing agent (precipitated calcium carbonate, etc.), fingerprints, and dust are present as contamination on polished surfaces of glasses subjected to a dimming and stain-removal treatment; whereas fingerprints, dust and dimming are present as contamination on polished surfaces of glasses that are not subjected to a dimming and staining removal treatment.

Contamination found on polished surfaces is classified according to components in Table 4.5. There are two primary sources of contamination: external factors, i.e., polishing agent, glass-fixing pitch, dust, etc.; and contamination arising from the optical glasses themselves in the case of poor chemical durability, i.e., dimming.

How does contamination adhere to the glass surface? Organic contaminants such as fingerprints, grinding oil, pitch, protective film material, etc., may apparently adhere to the polished surface either by simple physical adsorption, or through the formation of hydrogen bonds between polar groups in the organic substances and OH groups on the glass surface.[28]

Dust contaminants float through the air and consist of small particles such as dust from personal clothing and dust generated from soil, etc. These particles apparently adhere to the glass surface through the following mechanisms:

1.   Oil vapor in the air may act as an adhesive and cause particles to adhere to the glass surface.

2.   Water molecules on the surfaces of the particles may form hydrogen bonds with OH groups on the glass surface.

3.   Dimming may be generated at the interfaces between particles and the glass; this may act as an adhesive and cause the particles to adhere to the glass surface.

4.   Particles may adhere to the glass surface through electrostatic attraction.

5.   Particles may adhere to the glass surface through van der Waals force.

When contamination occurs from the polishing agent, some of the polishing agent that was not washed away by water immediately after polishing may have remained embedded in the glass surface.

Dimming is surface deterioration caused by a reaction between glass components and small amounts of water adhering to the glass surface.

Contaminants that occur on polished surfaces of optical glasses are shown schematically in Fig. 4.81.

Table 4.5.   Contamination on polished glass surfaces (by component).

Organic substances	Oil form	Fingerprints (oil portion), grinding oil, oil mist
	Solid	pitch, adhesive, protective film
Inorganic substances	Water soluble (salts)	Dimming, fingerprints (salt portion)
	Water insoluble (particles)	Polishing agent, dust

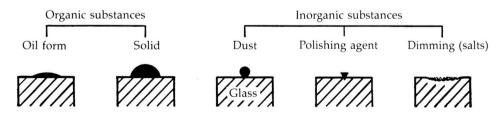

Figure 4.81.   Adhesion of contaminants to polished surfaces of optical glasses.

4.5.2. How Can Contaminants Be Removed?

The cleaning of optical glasses is accomplished by physical stripping actions such as hand wiping, brushing, ultrasonic cleaning, etc., chemical dissolving actions by water, organic solvents, synthetic detergents (surfactants plus inorganic builders), etc., and composite actions consisting of combinations of the two types of actions.

4.5.2.1. Physical Stripping Actions

Hand Wiping. Hand wiping is a method in which oily contaminants or solid particulate contaminants adhering to the glass surface are wiped away with a lens tissue or cotton cloth impregnated with an organic solvent such as ethanol or acetone, etc. However, this method suffers from the following drawbacks: execution of the method requires skill, the work efficiency is poor, and the adhesive strength of vacuum-evaporated coatings is weak in the case of hand-wiped surfaces.

Ultrasonic Cleaning. In the ultrasonic cleaning method,[29,30] contaminants adhering to the glass surface are stripped away as a result of the following effects: minute vibration, the agitation of the liquid, and the cavitation in the liquid created by the application of ultrasonic vibrations in a liquid. Particularly in the case of cavitation, contaminants adhering to the surface of the object to be cleaned are destroyed and stripped away by a microscopic

destructive action that is based on the formation and explosive destruction of vacuum bubbles occurring when gas molecules dissolved in the liquid are subjected to the expansive and compressive forces created by ultrasonic vibrations (Fig. 4.82). However, a sufficient cleaning effect cannot be obtained by means of this ultrasonic action alone. Effective cleaning can be achieved by combining this ultrasonic action with the chemical dissolving action of a synthetic detergent, etc.

4.5.2.2. Chemical Dissolving Action

Water. Water is the most common and most important cleaning agent. It is indispensable for the removal of water-soluble contaminants by dissolution and the flushing of detergents and contaminants, and as a medium for water-soluble detergents. It is purified as necessary before use. Light dimming on a polished surface will readily dissolve in water. Furthermore, in the case of glasses with a poor chemical durability, the glass itself will dissolve.

Organic Solvents. Organic compounds such as trichloroethylene, perchloroethylene, etc., are used to remove pitch and protective film materials, which are organic solid contaminants. In general, an organic solvent itself has absolutely no erosive effect on the glass surface. However, unstable organic compounds may be decomposed by sunlight, heat, etc., and thus generate acids such

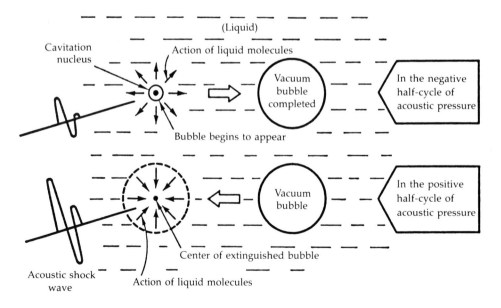

Figure 4.82. Model diagram of cavitation (Ref. 29).

as HCl, etc., that erode the glass. Consequently, care should be exercised in the use of such solvents.

Synthetic Detergents. Synthetic detergents[31] consist of surfactants and inorganic builders. Surfactant is a general name for substances that consist of hydrophobic groups and hydrophilic groups. These substances collect at interfaces between mutually immiscible substances such as oil and water, and thus lower the interfacial tension. In the generally used classification according to ionic type, surfactants are divided into four types (as shown in Fig. 4.83): nonionic, anionic, cationic, and ampholytic. Generally, however, only anionic and nonionic surfactants are used in synthetic detergents.

Surfactants dissolve in water by virtue of their hydrophilic groups. However, when the concentration of a surfactant exceeds the critical micelle concentration, several tens to several hundreds of molecules gather and form spherical, rod-form, or laminar micelles with the hydrophobic groups of the molecules facing inward, and the hydrophilic groups facing outward. Oils that do not readily dissolve in water are solubilized by being incorporated into the micelles. In addition to this solubilizing action, surfactants, by lowering the interfacial tension, possess wetting, penetrating, emulsifying, dispersing, bubble-generating, bubble-eliminating, and cleaning actions. Oily contaminants on polished surfaces are emulsified and dissolved away by the action of surfactants.

Inorganic builders are inorganic compounds that generally do not have any cleansing power themselves, but do have auxiliary powers that help the cleaning effect of surfactants when these builders are used together with surfactants. Commonly used examples are Na_2CO_3, Na_2SO_4, $Na_5P_3O_{10}$, $K_4P_2O_7$, and Na_2SiO_3.

Oily contaminants, such as fingerprints, can be removed by the emulsifying and solubilizing actions of a surfactant. On the other hand, inorganic contaminants that cannot be removed by dissolution (e.g., polishing agents) can be removed by the dissolution of the glass surface itself (to which the contaminants are adhering) using

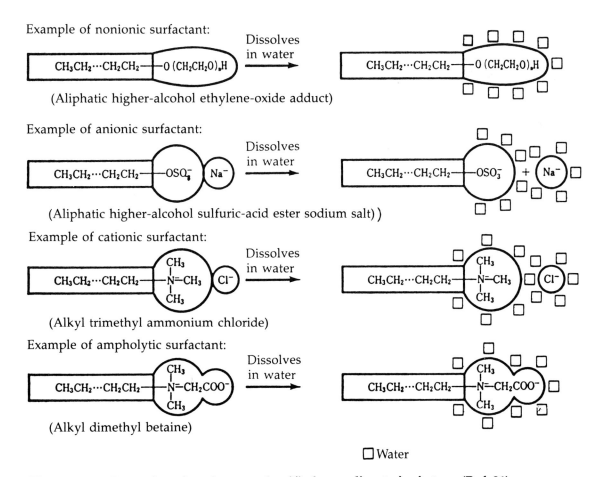

Example of nonionic surfactant:

(Aliphatic higher-alcohol ethylene-oxide adduct)

Example of anionic surfactant:

(Aliphatic higher-alcohol sulfuric-acid ester sodium salt))

Example of cationic surfactant:

(Alkyl trimethyl ammonium chloride)

Example of ampholytic surfactant:

(Alkyl dimethyl betaine)

□ Water

Figure 4.83.   Examples of surfactants classified according to ionic type (Ref. 31).

inorganic builders. In cleaning optical glasses, inorganic builders play an extremely important role (unlike their usual role as auxiliaries to surfactants) by eroding and dissolving the glass surface to which contaminants are adhering, and thus removing the contaminants from the glass surface so that the surface is cleaned.

4.5.2.3. Mechanisms of Contaminant Removal

Figure 4.84 shows various mechanisms of contaminant removal. Among organic contaminants, contaminants such as fingerprints, etc., can be removed by dissolution by an organic solvent, or by the emulsifying and solubilizing actions of a surfactant. Furthermore, solid organic contaminants such as pitch, protective film materials, etc., can be removed by dissolution by an organic solvent such as trichloroethylene, etc.

Among inorganic contaminants, dimming can be removed by dissolution by water. In the case of severe dimming, however, the roughness of the glass surface may progress to a point where clouding is generated.

Insoluble particulate contaminants such as polishing agents, dust, etc., are removed either by stripping the contaminants from the glass by physical force (hand wiping, ultrasonic cleaning, etc.), or by dissolving the glass surface.

4.5.2.4. Cleaning Apparatus

A multitank, ultrasonic, automatic-cleaning apparatus that combines the physical stripping action of ultrasonic waves with the chemical dissolving action of a synthetic detergent is generally used in the cleaning of optical glasses. Figure 4.85

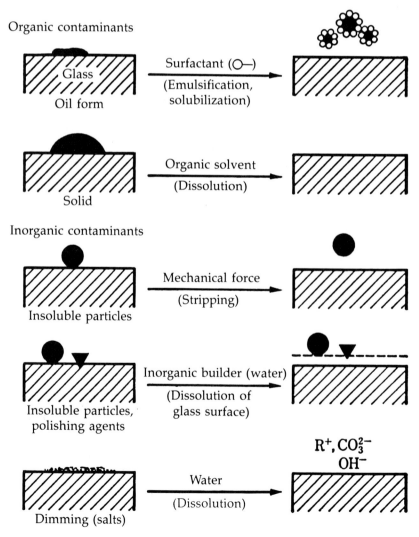

Figure 4.84.   Mechanisms of contaminant removal.

shows the combination of tanks in the most common type of ultrasonic automatic-cleaning apparatus, and the purpose of each tank.

In addition to the representative cleaning apparatus described above, an all-tank organic-solvent system,[32] a powder jet system, and a horn-type high-powered ultrasonic-cleaning system[33] also exist to prevent latent scratching. In the all-tank organic-solvent system, a surfactant is suspended in the organic solvent. In this system, organic contaminants are dissolved and removed by trichloroethylene, and water-soluble contaminants are dissolved and removed by the surfactant. No latent scratching occurs, but there is some difficulty in terms of the degree of cleaning. The horn-type high-powered ultrasonic-cleaning system uses a high-output (1.8 kW) horn-type oscillator; as a result, the cleaning time is reduced without any increase in the erosion rate. This fact is used to suppress the generation of latent scratches.

Another type of automatic cleaning apparatus has a plasma drying system,[34] which does not use flammable isophthalic acid (IPA) for safety reasons. In addition, a cleaning method exists in which water on the surface of the object being cleaned is removed using a freon tank containing a dehydrating agent instead of IPA.

In using a cleaning apparatus, it is necessary to gain a thorough understanding of the mechanical and chemical properties of the glass involved and of the properties of the synthetic detergent used. It is also necessary to exercise caution so that no staining or latent scratching is generated on the glass surface by the cleaning action.

4.5.2.5. Surface Deterioration in Optical Glasses during Cleaning—Staining and Latent Scratching

There are cases in which the operation of cleaning and removing contaminants from the glass surface actually generates faults in the glass

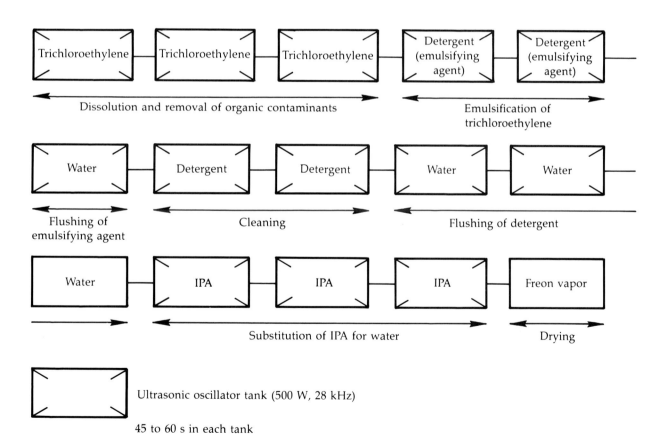

Figure 4.85.   Combination of tanks in ultrasonic automatic-cleaning apparatus.

surface. These faults are staining and latent scratching. The following has been observed with regard to staining. In the case of common silicate glasses, staining is generated by ion exchange between alkali components and water in the water tanks of the cleaning apparatus. In the case of lanthanum borate glasses, on the other hand, we observed the formation of a staining layer (consisting principally of La_2O_3) by the leaching of B_2O_3 and alkali-earth ions.

Latent scratching is generated by the erosive action of inorganic builders in the synthetic detergent tanks. For silicate glasses, when a polished surface is present, the surface is dissolved by OH^- ions generated by the hydrolysis of inorganic builders resulting in latent scratching. In the case of borate lanthanum glasses, latent scratching is generated by the dissolution of the lanthanum by $Na_5P_3O_{10}$, which is included as an inorganic builder.

Tables 4.6 and 4.7 show the relationship of immersion time to the generation of latent scratching and staining. This relationship was observed when samples of various optical glasses, which had been polished under a load of 100 g/cm^2 using polyurethane pads and 10 wt% CeO_2 suspended in water, were immersed in 16 liters of distilled water and synthetic detergent A-77 di-

luted to 1 vol% at 50°C and were subjected to ultrasonic waves (the load in the case of LaK 8, LaK 9^N, and LaK 14 was 135 g/cm^2).

Staining. For an immersion time of less than 20 minutes in distilled water, almost no optical glasses were stained (i.e., showed a brown interference color). Furthermore, staining was not observed at all in the case of immersion in the detergent solution. The reason for this appears to be that dissolution of the glass surface as a whole is the main reaction in a detergent. Among commercially marketed detergents, however, there are some detergents that promote selective leaching of components of the glass. Accordingly, caution should be exercised.

Latent Scratching. Among the 28 glasses tested, 11 glasses showed latent scratching at an immersion time of 10 minutes or less in distilled water, and 6 glasses showed latent scratching at an immersion time of 4 minutes or less in the detergent solution. In these cases, countless minute scratches caused the surfaces to exhibit a white cloudiness. In obtaining a highly clean surface, dissolution of the surface cannot be avoided; however, this is accompanied by surface deterioration: namely, the generation of latent scratches. Finishing of the glass to a scratch-free surface in the

Table 4.6.   Relationship between immersion time of glass in distilled water and generation of latent scratching and staining.

Immersion time (min)	Glasses with visible scratches using a slide-projector light	Glasses with visible brown staining
>60	FK 5, BK 7, K 3, BaK 4, BaSF 7, BaSF 8, BaSK 15, BaF 10, BaF 11, NbSF 13	All glasses except LaLK 6
20 to 30	SK 5, LaK 13, LaF 2, LaLF 3, LaLF 5, TaF 1	LaLK 6
10 to 20	SSK 5	
5 to 10	SK 16, SK 18, LaK 8, LaK 10	
<5	SK 4, SK 10, SK 14, LaLK 6, LaK 9, LaK 12, LaK 14	

Table 4.7.   Relationship between immersion time in a solution of detergent A-77 and generation of latent scratching and staining.

Immersion time (min)	Glasses with visible scratches using a slide-projector light	Glass with visible brown staining
>8	FK 5, BK 7, K 3, BaK 4, SK 5, SK 18, SSK 5, BaSF 7, BaSF 8, BaSF 15, BaF 10, BaF 11, LaK 8, LaK 10 LaF 2, LaLF 3, LaLF 5, NbSF 13, TaF 1	No staining observed in any of the glasses
4 to 8	SK 16, LaK 12, LaK 13	
2 to 4	SK 4, LaK 14	
<2	SK 10, SK 14, LaK 9, LaLK 6	

polishing process and the use of a low-erosion cleaning method in the cleaning process are effective in preventing latent scratching; however, the only fundamental solution to the problem is to improve the chemical durability of optical glasses.

4.5.3. Cleaning and Latent Scratching

Latent scratching first became a problem after the introduction of multilayer coatings. Latent scratching certainly existed prior to this time; it appears, however, that it was not a problem because dimming, clouding, contamination, and latent scratching on the lens surface were difficult to find with only a single-layer coating present.

It is debatable whether the Japanese word *senshō* (latent scratches) should be written with the characters *sen-shō* (latent scratch) or the characters *sen-shō* (washing scratch)! In view of the fact that the causes of such scratches occur in the polishing process, though the scratches themselves are latent at this time and therefore cannot be observed, it is justifiable to call such scratches "latent scratches." On the other hand, in view of the fact that such scratches can first be sensed and thus observed only after cleaning, we might also write the Japanese word as *senshō* (washing scratches). Here, for convenience, we will use the term latent scratches.

Latent scratches are first seen after cleaning in the synthetic detergent tanks. However, it is necessary to demonstrate clearly whether these scratches are created by cleaning, or whether the preceding polishing process has some connection with the scratches. Here, we will discuss the relationships between latent scratching and the polishing process, latent scratching and the cleaning process, and latent scratching and glass materials; in addition, we will try to clarify the true nature and causes of latent scratching. We will also analyze the cleaning process, clarify the relationship between synthetic detergents and latent scratching, point out the contradictions contained in automatic cleaning methods currently in use and discuss methods for the prevention of latent scratching.

4.5.3.1. Latent Scratching and the Polishing Process

To ascertain whether or not a relationship exists between the polishing process and latent scratching, we investigated the relationships between latent scratching and polishers, latent scratching and polishing agents, and latent scratching and the mechanical conditions of polishing. The glass used in this case was SSK 4. Water, a neutral detergent and hydrofluoric acid were used in the detergent tank during the experiments. The magnification of the photographs in this section is 200 ×. Cleaning was performed using the following tank sequence, shown in Fig. 4.86a.

Latent Scratching and Polishers. Figure 4.86b, c shows the results obtained when polytex and polyurethane pads are used as polishers. When polytex is used, no scratches are generated in SSK 4 by the water or synthetic detergent; when polyurethane pads are used, however, scratches are generated by the water and synthetic detergent. The difference is especially conspicuous when the glass is cleaned with hydrofluoric acid. This indicates that the generation of latent scratching is influenced by the properties of the polisher. Polytex is a somewhat elastic polisher, whereas polyurethane is a viscoelastic polisher. The latter is harder than the former. Thus, it seems that there is a great difference in the generation of latent scratches even under fixed polishing conditions.

Latent Scratching and Polishing Agents. The hardness of the polishing-agent cerium oxide increases with sintering temperature; BK 7 was polished using cerium oxide sintered at 1200°C and 900°C. As shown in Fig. 4.87, many more severe latent scratches were generated when cerium oxide sintered at 1200°C was used. Our interpretation is that the harder cerium oxide produced more scratches in the glass surface.

Latent Scratching and Mechanical Conditions of Polishing. The mechanical conditions of polishing are varied by varying the load, rotary

(a)

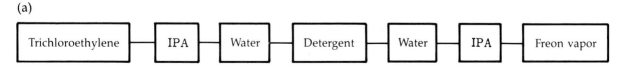

Figure 4.86a.   Tank sequence used for cleaning SSK 4 glass.

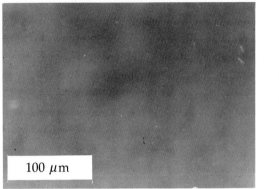

Water (120 s of exposure to ultrasonic waves)

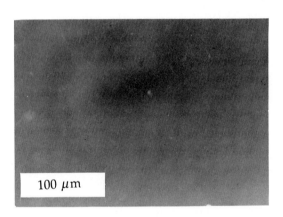

Detergent (40 s of exposure to ultrasonic waves)

Acid (40 s of exposure to ultrasonic waves)

(b) Polytex   (c) Polyurethane

Figure 4.86b. Polished surfaces (SSK 4) obtained using polytex as a polisher.

Figure 4.86c. Polished surfaces (SSK 4) obtained using polyurethane as a polisher.

speed, mechanism of the machine, etc. Figure 4.88, (a) shows a surface that was polished under conditions that resulted in a polishing rate approximately four times the polishing rate used in (b). Here, (a) shows more conspicuous scratches.

The results shown in Fig. 4.88 indicate that polishing conditions have an effect on latent

(a) Sintering temperature: 800°C   (b) Sintering temperature: 1200°C

Figure 4.87. Polished surfaces (BK 7) and sintering conditions of polishing agent (cerium oxide).

scratches generated after cleaning. It is shown that latent scratches will tend to occur if conditions are provided so that a hard substance scratches the glass surface. Latent scratches themselves cannot be observed in the polishing process. However, the germs of latent scratches are already formed in the polishing process; cleaning makes these scratches visible.

4.5.3.2. Latent Scratching and the Cleaning Process

In the lens cleaning process, pitch and protective film materials are dissolved by trichloroethylene, cerium oxide is removed by a synthetic detergent, and freon vapor is used for drying. Latent scratches are not generated in the case of organic solvents; they are formed in the synthetic detergent tank.

Erosion Characteristics and Cleaning Characteristics of Synthetic Detergents. Synthetic detergents generally consist of anionic or nonionic surfactants and inorganic builders. Accordingly, we tested the erosion characteristics of individual surfactants and inorganic builders with respect to

glasses and their cleaning capacities with respect to polishing agent particles and fingerprints. The results are shown in Tables 4.8 and 4.9. In the erosion test, the samples were immersed for one hour (with stirring) at 50°C, and the resulting weight loss was determined. Weight loss with respect to water was expressed as 1.

The erosion characteristics of surfactants are approximately the same as those of water. It appears that the surfactants themselves do not have an erosive effect on glasses; it is rather the water in which the surfactants are dissolved that erodes the glasses. Synthetic detergents have an erosive effect on glasses; however, the erosion characteristics in this case show roughly the same values as those of inorganic builders, indicating that the erosive effect of synthetic detergents arises from the inorganic builders. Accordingly, it appears that if the inorganic builders are removed from synthetic detergents, the detergents will cause no glass erosion, i.e., latent scratching.

In comparing the cleaning effects of surfactants, inorganic builders, and synthetic detergents, we adhered polishing agent particles or fingerprints to polished surfaces. These surfaces

(a) High-speed polishing   (b) Low-speed polishing

Figure 4.88.   Polished surfaces (BK 7) and mechanical conditions of polishing.

Table 4.8.   Erosion of glasses by cleaning liquids.

Cleaning liquid	pH	SSK 14	SK 16	LaLK 6	LaK 12	LaLF 3	
A-55	9.6	4.6	4.5	4.1	6.6	3.3	Anionic
A-77	11.9	3.8	1.9	2.2	2.2	0.7	Nonionic
Surfactant-40	6.5	1.9	1.7	1.7	2.1	1.2	Nonionic
Surfactant-95	6.5	2.7	2.0	1.1	1.1	0.9	Nonionic
Surfactant-END	6.5	1.6	2.9	—	—	—	Anionic
H_2O	6	1	1	1	1	1	
$Na_5P_3O_{10}$	8.7	4.1	7.7	16.1	36.8	7.6	
$NaHCO_3$	9	3.6	0.8	1.9	4.3	1.4	
$Na_4P_2O_7$	11	5.8	2.7	—	—	—	
Na_2CO_3	12	2.7	1.1	0.9	0.3	0.3	

were immersed in the various cleaning liquids, and we made a qualitative judgement of "effective" or "ineffective." As shown in Table 4.9, surfactants and synthetic detergents have a cleaning effect on fingerprints, but are ineffective in re-moving polishing agent particles. As for inorganic builders, it is seen that $Na_5P_3O_{10}$, which has the strongest erosive effect on glass, is effective in removing cerium oxide. This indicates that erosion of the glass is unavoidable if cerium oxide is to be

Table 4.9. Comparison of cleaning characteristics.

Cleaning agent	pH	SK 16		LaLK 6		
		Polishing agent	Fingerprints	Polishing agent	Fingerprints	
A-55	9.6	−	+	−	+	Anionic
A-77	11.9	−	+	−	+	Nonionic
Surfactant-40	6.5	−	+	−	+	Nonionic
Surfactant-95	6.5	−	+	−	+	Nonionic
Surfactant-END	6.5	−	+	−	+	Anionic
H_2O	6	−	−	−	−	
$Na_5P_3O_{10}$	8.7	±	−	±	−	
$NaHCO_3$	9	−	−	−	−	
$Na_4P_2O_7$	11	−	−	−	−	
Na_2CO_3	12	−	−	−	−	

successfully removed, or conversely, only substances that can erode the glass are able to remove cerium oxide.

Cleaning and Latent Scratching. If we assume that cleaning generally consists of the removal of substances adhering to a surface, then cleaning may be achieved either by dissolving the adhering substances or stripping them away. Even in the removal of cerium oxide from the glass surface, it appears that the CeO_2-glass interface is eroded by an acid or alkali to remove the cerium oxide. Accordingly, the reason that synthetic detergents have an effect in removing cerium oxide is that erosive inorganic builders are contained in these detergents. At the same time, however, inorganic builders also create latent scratches on the glass surface. In other words, if an attempt is made to remove cerium oxide, latent scratches are inevitably generated; cerium oxide cannot be removed without creating latent scratches. Thus, the removal of cerium oxide using a synthetic detergent is inevitably accompanied by the generation of latent scratches.

Surfactants cannot remove cerium oxide. Surfactants have lipophilic groups, and can therefore remove oils, but they do not have any affinity for inorganic substances such as cerium oxide, and therefore have no cleaning capacity in such a case. From this point as well, it is clear that surfactants cannot be used to remove cerium oxide.

4.5.3.3. Latent Scratching and Glass Materials

The cause of latent scratches lies in the scratching of the glass surface by foreign matter during the polishing process. As described above, these invisible scratches are enlarged by erosion during the cleaning process, and thus become visible. Therefore, a connection between latent

Table 4.10. Chemical durability and hardness of glasses.

	Chemical durability, D_W (class)	Hardness, H_K (kg/mm^2)
F 2	3	440
SSK 4	3	575
SK 16	5	570

scratching and physical properties of the glass involved is naturally inferred. In other words, the ease with which scratches are made is affected by the hardness of the glass, and the ease with which these scratches are enlarged depends on the chemical durability of the glass. Accordingly, we investigated the relationship between glass hardness and the tendency of latent scratching to occur, and the relationship between the chemical durability of glasses and the tendency of latent scratching to occur.

Latent Scratching and Glass Hardness. As shown in Table 4.10, SSK 4 and F 2 have approximately the same chemical durability, but F 2 has a lower hardness. A comparison of the results shown in Fig. 4.86b and Fig. 4.89 shows that latent scratches are more numerous in F 2.

Latent Scratching and Chemical Durability of Glasses. The glasses SSK 4 and SK 16 have approximately the same hardness, but SK 16 has a poorer chemical durability, i.e., class 5. As seen from a comparision of Fig. 4.86b and Fig. 4.90, SK 16, with poorer chemical durability, shows more latent scratches. Thus, latent scratching is affected by the hardness and the chemical durability of the glass involved. To decrease latent scratching, it is necessary to increase the hardness and improve the chemical durability of the glass.

Water (120 s of exposure to ultrasonic waves)

Water (120 s of exposure to ultrasonic waves)

Detergent (40 s of exposure
to ultrasonic waves)

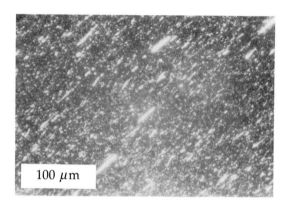

Detergent (40 s of exposure
to ultrasonic waves)

Acid (40 s of exposure to ultrasonic waves)

Acid (40 s of exposure to ultrasonic waves)

Figure 4.89. Polished surfaces of F 2 obtained using polyurethane as a polisher.

Figure 4.90. Polished surfaces of SK 16 obtained using polyurethane as a polisher.

In conclusion, latent scratches are minute scratches that are formed in the polishing process and are enlarged in the cleaning process so that they become visible. The cause of this phenomenon appears to be scratching during polishing, especially scratching by relatively large particles of cerium oxide or dust. The size of latent scratches is about 3 to 4 μm; thus, these scratches are essentially different from the approximately 80-Å scratches connected with the polishing mechanism.

The following three things should be done to eliminate latent scratching:

1. Examination of polishing conditions: soft polishing conditions, i.e., a soft polisher, low-hardness cerium oxide, a small load, etc., are desirable.

2. Examination of the cleaning process: methods that attempt to remove cerium oxide using a synthetic detergent will inevitably produce latent scratching. Although not offering a complete solution, cleaning by means of hand wiping, by an organic solvent, or by reducing the cleaning time by increasing the ultrasonic output should be considered in the case of glasses that are susceptible to latent scratching.

3. Improvement of glass materials: prevention of latent scratching from the standpoint of glass materials depends on improvement of the hardness and chemical durability of glasses. This is the only way that complete prevention of latent scratching will ever be achieved.

4.5.4. Summary

The purpose of cleaning is to remove contaminants from the surface of the object being cleaned so that a surface with the necessary degree of cleanness can be obtained in subsequent processes. Methods for obtaining a clean surface exist that perform the following functions:

1. Prevent the adhesion of contaminants in the first place.

2. Remove adhering contaminants or substances binding these contaminants by chemical dissolution or stripping.

3. Separate and remove contaminants from the surface of the object being cleaned by dissolving the surface without causing any abnormalities in the surface.

Figure 4.91 summarizes the process of cleaning optical glasses using an ultrasonic automatic

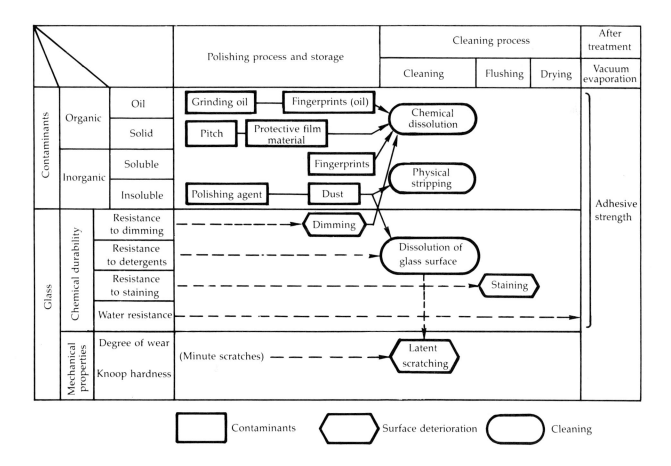

Figure 4.91.   Types of contaminants, properties of glasses, and cleaning.

cleaning apparatus. The adhesion of contaminants such as pitch, protective film materials, etc., during the polishing process and storage is unavoidable because of the manufacturing process used. However, care must be taken to avoid the adhesion or generation of contaminants such as fingerprints, dust, dimming, etc.

Hand washing and wiping immediately after polishing should produce an effective cleaning action (as might be expected from a hand-wiping operation); the thorough washing away of polishing-agent particles adhering to the surface in this stage should facilitate subsequent cleaning operations.

In cleaning, contaminants (e.g., grinding oil, pitch, protective film materials, fingerprints, dimming) that are soluble in solvents (water, surfactants, organic solvents) can be removed by chemical dissolution. Insoluble substances and contaminants that will not dissolve in solvents must be separated and removed from the surface by the additional use of ultrasonic waves or dissolution of the glass surface.

In some cases, small scratches generated in polishing grow because of dissolution of the glass surface, thus making them observable as latent scratches. To prevent the generation of latent scratches, it is necessary to make special efforts to obtain a scratch-free surface in polishing.

In the flushing process, the synthetic detergent used in cleaning and the contaminants removed from the surface are thoroughly washed away. The water used in flushing must be sufficiently clean to ensure that the contamination-free clean surface obtained by cleaning is not contaminated by the flushing water.

Some optical glasses have a poor chemical durability so that selective leaching of soluble components may occur during flushing, and thus lead to the generation of staining. Accordingly, surfaces should not be flushed by immersion in water for any longer than is necessary.

Drying ordinarily consists of vapor drying using an organic solvent. After the water used in flushing is replaced by a solvent, which is soluble in the organic solvent (freon) used in vapor drying, the cleaning process is completed by drying in such a way that the surface is not contaminated by removal of the solvent, etc.

Chemical reactions play an important role in the cleaning of optical glasses. It is important that cleaning be performed with a full understanding of the types of contaminants to be removed, the chemical properties of the glass being cleaned, and the characteristics of the cleaning method used.

4.6. Coating

In most cases, lenses and prisms are coated after being polished and cleaned. The current view is that vacuum-evaporated coatings adhere to the glass substrate by van der Waals force.[35-37] However, as described later, it has been found that there is a close relationship between the vacuum-evaporated coating and the substrate glass. Accordingly, it appears that there is a need to reexamine the manner of bonding between a vacuum-evaporated coating and the substrate glass.

In this section, we will discuss the following:
1. The effect that the cleaning of the substrate glass has on the adhesive strength of coatings.
2. The effect that surface deterioration of the substrate glass (dimming, staining, latent scratching, etc.) has on the adhesive strength of coatings.

3. The relationship between the chemical durability of the substrate glass and the adhesive strength of coatings.
4. The relationship between coatings and the composition of the substrate glass.

4.6.1. Relationship between Glass Surface Conditions and Adhesion of Vacuum-Evaporated Coatings

In this section, we discuss the results obtained when the relationship between glass surface conditions and the adhesion of coatings was investigated using the optical glasses BK 7, SF 6, SK 16, LaK 12, and NbFD 13.[38] The mechanical and chemical properties of these substrate glasses are shown in Table 4.11.

The glass BK 7 is an SiO_2–B_2O_3–R_2O glass and has a strong chemical durability. The glass SF 6 contains large amounts of PbO and is mechanically soft; dimming does not readily occur, but the chemical durability is poor. The glass SK 16 is an SiO_2–B_2O_3–BaO glass and has the poorest chemical durability. The glass LaK 12 is an SiO_2–B_2O_3–La_2O_3–RO glass and has a poor chemical durability; whereas NbSF 13 consists principally of B_2O_3, La_2O_3, and rare-earth elements and has a superior chemical durability.

After polishing, the samples were washed thoroughly with hot water and wiped with a cloth impregnated with ethyl ether. After the surface conditions were inspected, the samples were coated with a protective film, de-blocked from the block holder, and stored in a desiccator. The glass samples stored in the desiccator were taken out prior to vacuum evaporation and cleaned by several methods. The samples were then subjected to a vacuum evaporation treatment. The vacuum evaporation conditions are shown in Table 4.12.

The adhesive strength was measured by the scratching test method[39] that was proposed by Heavens,[40] and for which an analysis method was indicated by Benjamin et al.[41] A model diagram of the apparatus and the analysis method are shown in Figs. 4.92 and 4.93. A diamond indenter with the tip finished to a spherical shape (30-mm radius of curvature) was pressed perpendicularly

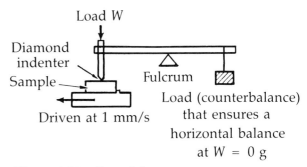

Figure 4.92.   Scratching-test apparatus.

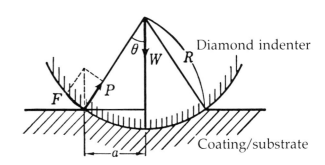

Figure 4.93.   Scratching-test analysis method.

Table 4.11.   Properties of substrate glasses.

| | Mechanical properties | | Chemical properties | | | | |
| | | | | Resistance to dimming, | Resistance to staining, | Resistance to detergents | |
Glass type	Degree of wear, F_A	Knoop hardness, H_K	Water resistance, D_0[a]	D_W[b]	T_{Blue}[c]	D_{NaOH}[d]	D_{STPP}[e]
BK 7	100	595	0.3	0.21	70	0.06	0.01
SF 6	230	365	11.2	0.02	—	0.16	0.26
SK 16	140	560	23.4	0.27	—	0.71	0.89
LaK 12	150	590	47.8	0.14	1	0.15	2.78
NbSF 13	70	680	1.4	0.02	45	0.01	0.02

[a] Weight loss (10^{-3} mg/cm^2·h).
[b] Weight loss (wt%).
[c] Time required for interference color to change from brown to blue (h).
[d] Weight loss (mg/cm^2·15 h).
[e] Weight loss (mg/cm^2·h).

Table 4.12.   Vacuum evaporation conditions.

	Heating method	Substrate temperature (°C)	Degree of vacuum (Torr)	Vacuum evaporation rate (Å/s)	Coating thickness (Å)
MgF_2	Tungsten coat resistance heating	250	3×10^{-5}	3 to 10	900
Al_2O_3	Electron beam heating	250	3×10^{-5}	3 to 10	2200

against the glass surface coated with a vacuum evaporation coating; with a load applied, the surface was scratched at a constant rate. The shearing force F was determined from the load at which the coating began to strip away from the edges of the scratch, and the adhesive strength was evaluated in these terms.

$$F = P \tan \theta = \frac{a}{\sqrt{R^2 - a^2}} P \ . \qquad (4.2)$$

$$P = W/\pi a^2 \ . \qquad (4.3)$$

Here, W is the load at which the coating begins to strip away, R is the radius of curvature of the indenter, a is the radius of the indentation, and P is the hardness of the glass.

The four cleaning methods shown in Fig. 4.94 were examined to determine the effect of cleaning on adhesive strength. The expected cleaning effects are as follows:

• Cleaning method I is a common method that uses a synthetic detergent (A-77,* diluted to 1 vol%); the cleanest surface is obtained using this method. Organic contaminants and polishing agent particles are removed, but latent scratching occurs.

• Cleaning method II is a method which uses only a surfactant [Nonipole 95[†] (transliteration of a trade name) diluted to 0.01 vol%]. In this case, no latent scratches or organic contaminants are found on the cleaned surface, but polishing agent particles cannot be removed because the cleaning liquid contains no inorganic builder.

_____

* Shimada Rika Kogyo (K.K.)
[†] Sanyo Kasei Kogyo (K.K.)

• Cleaning method III consists of cleaning by distilled water alone; there is no latent scratching, but organic contaminants and polishing agent particles are not removed.

• Cleaning method IV consists of cleaning by hand wiping with a cloth impregnated with ethyl ether. With this method, there is no latent scratching, and there should be no organic contaminants or polishing agent particles present on the cleaned surface; however, the cleaning appears to be imperfect.

Staining, dimming, and latent scratching were deliberately induced on the glass surface to confirm the effect of surface deterioration on the adhesive strength of coatings. Tests were also conducted with contaminants such as fingerprints and CeO_2 polishing-agent particles applied to the surface. The experimental results are shown in Fig. 4.95 (MgF_2 coatings) and Fig. 4.96 (Al_2O_3 coatings).

4.6.1.1. Relationship between Cleaning Method and Adhesive Strength of Coatings

In the case of MgF_2 coatings, almost no difference exists between cleaning method I (synthetic detergent) and cleaning method II (surfactant). This probably indicates that the presence of latent scratches has no effect on the adhesive strength of coatings. The adhesive strength of coatings on surfaces treated by cleaning method III (water) was inferior to the adhesive strength of coatings on surfaces treated by cleaning methods I and II (using a synthetic detergent and a surfactant, respectively). The reason for this is probably that organic contaminants (pitch, protective film materials, fingerprints, etc.) and CeO_2 polishing agent particles could not be completely removed in the case of method III. The adhesive strength of

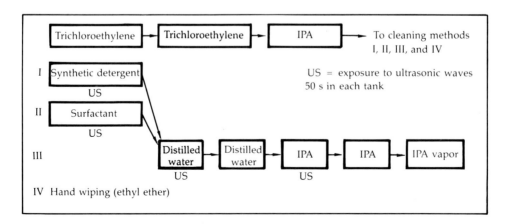

Figure 4.94. Substrate cleaning methods.

coatings on surfaces treated by cleaning method IV, i.e., hand wiping using a cloth and ethyl ether, was inferior to that observed in the case of methods I and II (using a detergent and surfactant). The reason for this appears to be that oily substances eluted from the cloth or fingers during hand wip-

ing form a thin adhering film on the glass surface, and the vacuum-evaporated coating is applied on top of this.

In regard to the adhesive strength of Al_2O_3 coating, the cleaning method used had no effect in any of the glasses tested. The reason for this is

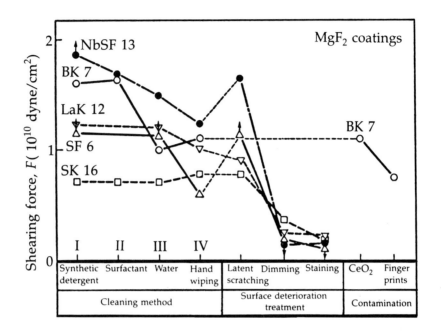

Figure 4.95. Adhesive strength of vacuum-evaporated MgF_2 coatings.

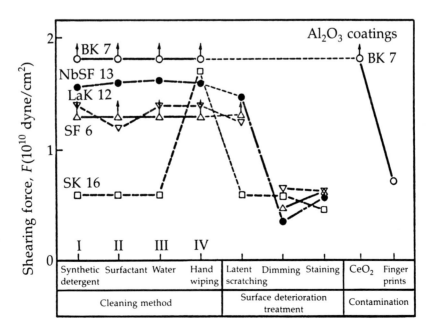

Figure 4.96. Adhesive strength of vacuum-evaporated Al_2O_3 coatings.

thought to be that it is difficult for surface treatments to show any effect because the adhesive strength of an Al_2O_3 coating is greater than that of an MgF_2 coating. On the other hand, the lack of any variation in the adhesive strength of coatings to SK 16 in cleaning methods I, II, and III probably results from the formation of weak staining during the flushing process, with the coatings being vacuum evaporated on top of this staining. The adhesive strength seen here is approximately the same as the adhesive strength of coatings to staining layers described later, with low values being shown. In the case of hand wiping, it appears that no staining is generated during cleaning, and that staining which was already present is removed by the wiping so that the adhesive strength of the Al_2O_3 coating to the glass itself is measured, and high values are obtained.

4.6.1.2. Relationship between Surface Deterioration and Adhesive Strength of Coatings

As shown in Figs. 4.95 and 4.96, countless minute scratches (latent scratches) generated on the glass surface do not cause any lowering of the adhesive strength of either MgF_2 or Al_2O_3 coatings. In regard to dimming, the adhesive strength of both MgF_2 coatings and Al_2O_3 coatings to a glass surface where dimming is present is greatly lowered in all the glasses tested. The reason for this is probably that peeling occurs between the glass and the dimming layer.

In the case of staining, the adhesive strength of coatings to a surface where staining is present is extremely low. The reason for this is probably that peeling between the substrate glass and the staining layer (and not peeling between the surface of the staining layer and the coating) is measured in the scratching test. The staining layer is a water-containing framework of network-forming oxides that is formed by ion exchange between soluble cations in the glass surface and hydronium ions in water. Because this layer is more porous and mechanically weaker than the glass matrix, it appears that the staining layer easily peels away from the glass matrix, and thus weakens the adhesive strength of the coating.

4.6.1.3. Relationship between Contamination and Adhesive Strength of Coatings

The adhesion of insoluble inorganic contaminants to the glass surface, e.g., the adhesion of cerium oxide polishing agent particles to a BK 7 surface, lowers the adhesive strength of an MgF_2 coating. In the case of a BK 7 surface contaminated with organic contaminants, e.g., fingerprints, both MgF_2 and Al_2O_3 coatings show a decrease in adhesive strength. It appears that when fingerprints adhere to a glass surface, the surface energy of the substrate drops; as a result, the adhesive strength of the coating drops, and the coating becomes susceptible to peeling. The low adhesive strength found for surfaces cleaned by hand wiping is probably also related to this phenomenon. Thus, dimming, staining, and contamination not only lower the apparent value of the glass, but also lower the adhesive strength of coatings. Sufficient attention must be paid to the prevention of dimming and staining, and to the cleaning and removal of contaminants.

4.6.2. Structure of Staining

The staining layer is a thin film that is forced as a result of the selective leaching of glass components (alkali, etc.).[42] The structure of this layer can be expressed by a heterogeneous model[43-45] with an index-of-refraction slope such as those shown in Fig. 4.97A. The index of refraction n_{fa} of the face of the staining layer that is in contact with the air, the index of refraction n_{fg} of the face of the staining layer that is in contact with the glass matrix, and the thickness d_f of the staining layer can be determined from the maximum and minimum of the spectroscopic reflectance curve using the following equations:

$$\left(n_{fa} + n_{fg} \right) d_f = \frac{2m\lambda}{2} \quad ,$$

$$R\,\text{max} = \frac{\left(n_{fa}n_g - n_{fg} \right)^2}{\left(n_{fa}n_g + n_{fg} \right)^2} \quad ;$$

$$\tag{4.4}$$

$$\left(n_{fa} + n_{fg} \right) d_f = \frac{(2m + 1)\lambda}{2} \quad ,$$

$$R\,\text{min} = \frac{\left(n_{fa}n_{fg} - n_g \right)^2}{\left(n_{fa}n_{fg} + n_g \right)^2} \quad .$$

Here, m is an integer and n_g is the index of refraction of the substrate glass.

For the thin film structures a, b, and c in Fig. 4.97A, the corresponding spectroscopic reflectance curves a, b, and c are obtained (as shown in Fig. 4.97B).

Using borate glasses, silicate glasses, and phosphate glasses of the simple compositions shown in Table 4.13(A), (B), and (C), staining was

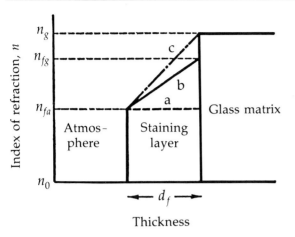

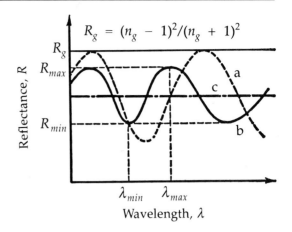

(A) Models of index-of-refraction distributions in staining layers

a: Has a uniform index of refraction

b: $n_g \neq n_{fg} \neq n_{fa}$ heterogeneous film

c: $n_g = n_{fg} \neq n_{fa}$ heterogeneous film

(B) Spectroscopic reflectance curves of staining layers

R_g: Index of refraction of the glass

Figure 4.97. Models of staining layers with linear index-of-refraction distributions.

Table 4.13. Compositions of glasses (mol%) used to investigate the properties of staining.

(A) Borate glasses

	B_2O_3/BaO					La_2O_3/BaO			RO				B_2O_3/SiO_2		
	B2	1	3	4	5	B1	6	7	B8	9	10	11	B12	13	14
B_2O_3	80	75	70	65	60	75	75	75	57.5	57.5	57.5	57.5	70	62.0	55
SiO_2	—	—	—	—	—	—	—	—	—	—	—	—	5	12.5	20
La_2O_3	5	5	5	5	5	5	10	15	7.5	7.5	7.5	7.5	5	5	5
RO	15	20	25	30	35	20	15	10	35.0	35.0	35.0	35.0	20	20	20
	Ba	Ba	Ba	Ba	Ba	Ba	Ba	Ba	Mg	Ca	Sr	Ba	Ba	Ba	Ba

(B) Silicate glasses

	SiO_2/Na_2O			Al_2O_3/BaO			R_2O			RO			
	S1	2	10	S3	4	11	S5	3	6	S7	8	9	3
SiO_2	65	75	85	65	65	65	65	65	65	65	65	65	65
Al_2O_3	—	—	—	0	5	10	—	—	—	—	—	—	—
R_2O	35	25	15	15	15	15	15	15	15	15	15	15	15
	Na	Na	Na	Na	Na	Na	Li	Na	K	Na	Na	Na	Na
RO	—	—	—	20	15	10	20	20	20	20	20	20	20
				Ba	Ba	Ba	Ba	Ba	Ba	Mg	Ca	Sr	Ba

(C) Phosphate glasses

	P_2O_5/Na_2O			Al_2O_3/CaO			R_2O			RO			
	P2	1	3	P9	10	11	P4	1	5	P6	1	7	8
P_2O_3	50	55	65	60	60	60	55	55	55	55	55	55	55
Al_2O_3	5	5	5	5	10	15	5	5	5	5	5	5	5
R_2O	25	20	10	20	20	20	20	20	20	20	20	20	20
	Na	Na	Na	Na	Na	Na	Li	Na	K	Na	Na	Na	Na
RO	20	20	20	15	10	5	20	20	20	20	20	20	20
	Ca	Ca	Ca	Ca	Ca	Ca	Ca	Ca	Ca	Mg	Ca	Sr	Ba

generated by immersing the glasses in pure water at 50°C, and the structures of the staining layers were investigated by the method described previously. Here, we will discuss the results. The index-of-refraction distributions of the staining layers in the borate glasses, silicate glasses, and phosphate glasses can be respectively expressed as in (a), (b), and (c) in Fig. 4.98.

4.6.2.1. Borate Glasses

In the case of borate glasses, B_2O_3 and BaO are almost completely leached from the staining layer. As a result, the principal component of the staining layer is La_2O_3, and the staining layer formed is porous and fragile. This staining layer is observed in glasses with a low water-resistance weight loss D_0; the structure of the layer appears to be as shown in Fig. 4.98a. Where n_g of the glass matrix is ~1.65, n_f of the staining layer is ~1.40, and the layer shows an index-of-refraction slope of $\Delta n = n_{fg} - n_{fa} \sim 0.06$. For example, in the case of the 57.5% B_2O_3, 7.5% La_2O_3, 35.0% SrO (mol%) glass, where $n_g = 1.67$, a sample that was immersed for five hours showed the following staining: $n_{fg} = 1.42$, $n_{fa} = 1.36$, and $d_f = 970$ nm. Electron micrographs of the surface conditions are shown in Fig. 4.99. This figure shows that the surface becomes rougher with longer immersion time. Furthermore, in glasses with a high D_0, the rate of staining formation is rapid, and collapse resulting from internal stress is common.

4.6.2.2. Silicate Glasses

In the case of silicate glasses, glasses with a high water-resistance weight loss D_0 show the formation of a thick staining layer ($d_f >$ several μm), and cracking tends to occur. Furthermore, an extremely rough staining layer is formed. Where D_0 is low, a strong staining layer that shows a uniform interference color is formed. The structure of the layer is shown in Fig. 4.98b. At an n_g of ~1.57, n_f of the staining layer is ~1.48; as the layer grows thicker, an index-of-refraction slope Δn of ~0.04 is shown. For example, in the 65% SiO_2, 15% Na_2O, 20% SrO (mol%) glass, where $n_g = 1.55$, a sample that was immersed for 24 hours showed the following staining: $n_{fg} = 1.51$, $n_{fa} = 1.47$, and $d_f = 870$ nm. Electron micrographs of the surface are shown in Fig. 4.100. This figure shows that the surface grows rougher with time, as in the case of borate glasses.

4.6.2.3. Phosphate Glasses

In phosphate glasses, the glass as a whole is dissolved so that no staining layer is formed re-

gardless of the value of the water-resistance weight loss D_0. Accordingly, the surface structure can be expressed by $n_g = 1.52$ alone, as shown in Fig. 4.98c. Electron micrographs of 55% P_2O_5, 5% Al_2O_3, 20% Na_2O, 20% CaO (mol%) glass surfaces are shown in Fig. 4.101. Unlike the case of borate glasses or silicate glasses, scratches that were initially present are caused to fade by the dissolution of the glass as a whole; after 34 hours, these scratches are completely removed so that a smooth surface is obtained.

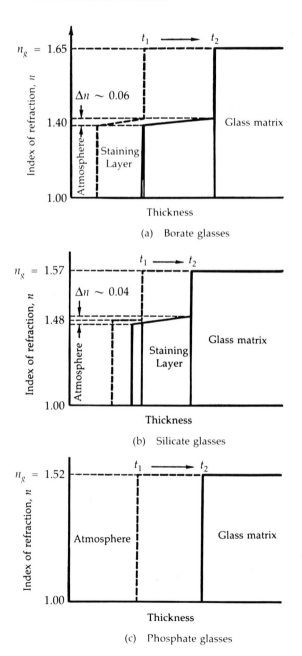

(a)   Borate glasses

(b)   Silicate glasses

(c)   Phosphate glasses

Figure 4.98. Index-of-refraction distributions of staining layers determined from spectroscopic reflectance curves.

(a) Polished surface

(b) Surface after 2 hr of immersion in pure water

(c) Surface after 5 hr of immersion

Figure 4.99. Transmission electron micrographs of borate glass surfaces (57.5 mol% B_2O_3, 7.5 mol% La_2O_3, 35 mol% SrO) ($\times$ 30,000).

(a) Polished surface

(b) Surface after 10 hr of immersion in pure water

(c) Surface after 24 hr of immersion

Figure 4.100. Transmission electron micrographs of silicate glass surfaces (65 mol% SiO_2, 15 mol% Na_2O, 20 mol% SrO) ($\times$ 30,000).

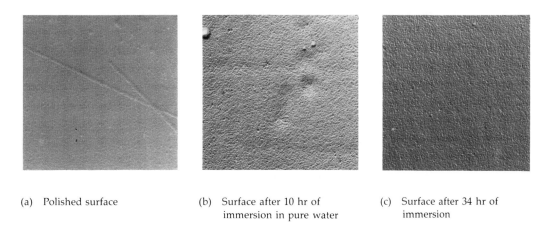

(a) Polished surface

(b) Surface after 10 hr of immersion in pure water

(c) Surface after 34 hr of immersion

Figure 4.101. Transmission electron micrographs of phosphate glass surfaces (55 mol% P_2O_5, 5 mol% Al_2O_3, 20 mol% Na_2O, 20 mol% CaO) ($\times$ 30,000).

4.6.3. Water Resistance of Glasses and Adhesive Strength of Vacuum-Evaporated Coatings

Figure 4.102a, b, and c shows the relationship between glass water-resistance weight loss D_0 and adhesion characteristics of vacuum-evaporated coatings for the borate, silicate, and phosphate glasses (the compositions are shown in Table 4.13). The vacuum evaporation conditions are shown in Table 4.14. Furthermore, adhesive strength was measured by the aforementioned scratching test.

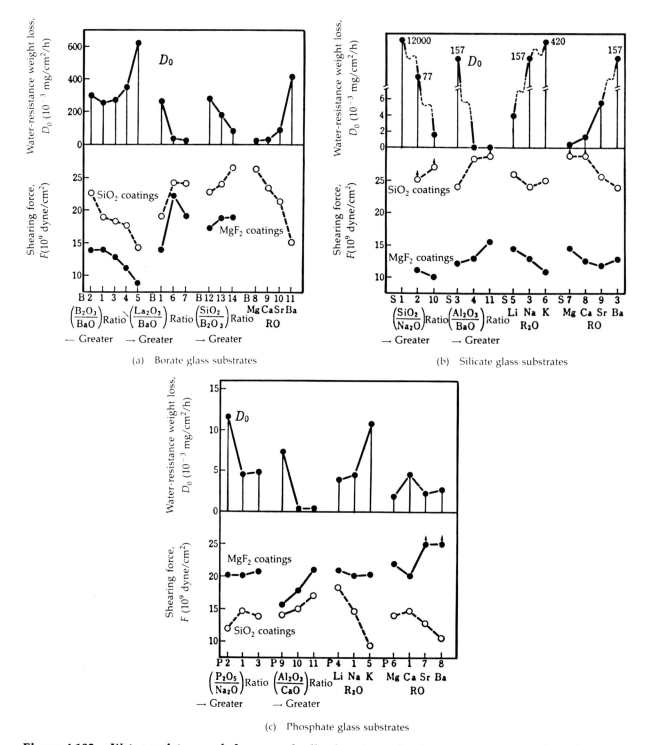

(a)   Borate glass substrates

(b)   Silicate glass substrates

(c)   Phosphate glass substrates

Figure 4.102.   Water resistance of glasses and adhesive strength of vacuum-evaporated coatings.

4.6.3.1. Borate Glasses

In borate glasses, the replacement of B_2O_3 by BaO results in a deterioration of the water resistance and a decrease in the adhesive strength of MgF_2 coatings and SiO_2 coatings. Furthermore, when the amount of La_2O_3 is increased, the water resistance is improved and the adhesive strength of the coatings increases. When B_2O_3 is replaced by SiO_2, the water-resistance weight loss D_0 decreases and the adhesive strength increases. In the case of divalent components, the water-resistance weight loss increases and the adhesive strength of coatings decreases with a decrease in the cationic field strength z/a^2. Thus, there appears to be a correlation here: the adhesive strength of coatings decreased with a decrease in the chemical durability of the glass. However, it is uncertain whether this is just an apparent correlation resulting from both properties being governed by cationic field strength, or whether the adhesive strength actually is governed by the chemical durability or ease with which a staining layer is formed. In the case of borate glasses, furthermore, it was found that SiO_2 coatings have a greater adhesive strength than MgF_2 coatings.

4.6.3.2. Silicate Glasses

A situation similar to borate glasses was observed for silicate glasses. As the modifying component Na_2O increases, the water-resistance weight loss D_0 increases and the adhesive strength of coatings decreases. When Al_2O_3 is introduced, the water resistance is improved and the adhesive strength of coatings increases. In regard to alkali and alkali-earth elements, the chemical durability improves and the adhesive strength of coatings increases with an increase in cationic field strength. Furthermore, in silicate glasses as well, SiO_2 coatings showed a greater adhesive strength than MgF_2 coatings.

4.6.3.3. Phosphate Glasses

Like borate and silicate glasses, a similar relationship was also seen in the case of phosphate glasses. As P_2O_5 increases, the water-resistance weight loss D_0 decreases and the adhesive strength of coatings tends to increase, though only slightly. As Al_2O_3 increases, the water resistance is improved and the adhesive strength increases. As the cationic field strength of alkali ions increases, the water-resistance weight loss decreases and the adhesive strength of coatings increases. In the case of alkali-earth ions, the tendency seen here differs from that observed in borate glasses and silicate glasses: namely, no fixed tendency is shown with respect to cationic field strength. However, a correspondence between chemical durability and adhesive strength is observed in the case of MgF_2 coatings. In phosphate glasses, furthermore, MgF_2 coatings show a greater adhesive strength than SiO_2 coatings.

4.6.4. Relationship between Glass Composition and Adhesive Strength of Coatings

As described previously, staining is not formed in water in the case of phosphate glasses. Accordingly, the data on adhesive strength in this case would appear to represent the actual adhesive strength between the glass and the vacuum-evaporated coating, independent of chemical durability (water resistance). The fact that the adhesive strength of a given coating increases with an increase in the cationic field strength of the glass would seem to suggest the presence of an electrostatic bonding force rather than van der Waals force between the coating and the substrate glass. Furthermore, oxide coatings were found to have a greater adhesive strength in the case of borate and silicate glasses, and fluoride coatings were found to have a greater adhesive strength in the case of phosphate (fluorophosphate) glasses.[47] The fact that there is thus a selectivity of coatings with good adhesion characteristics for various glass-forming oxides would appear to indicate that the bonds between the coating and the

Table 4.14. Vacuum evaporation conditions.

	Heating method	Substrate temperature (°C)	Degree of vacuum (Torr)	Vacuum evaporation rate (Å/s)	Coating thickness (Å)
MgF_2	Tungsten coat resistance heating	330	3×10^{-5}	3 to 10	2700
SiO_2	Electron beam heating	330	3×10^{-5}	3 to 10	2500

substrate glass are not only simple physical van der Waals bonds, but also chemical bonds. However, this remains to be confirmed by further study. I believe that it is necessary in the future to clarify the formation mechanism of vacuum-evaporated coatings on a glass surface and the adhesive strength between vacuum-evaporated coatings and glass components with external factors such as the formation of a staining layer excluded.

References

1. O. Imanaka, *Sci. Mach.* **20**, 60, 357 (1968).
2. L. Holland, *The Properties of Glass Surface*, Chapman & Hall, London, 1964).
3. W. F. Koehler, *J. Opt. Soc. Am.* **43**, 743 (1955); and W. F. Koehler and W. C. White, *J. Opt. Soc. Am.* **45**, 1015 (1955).
4. G. O. Rawstron, *J. Soc. Glass Technol.* **42**, 253T (1958).
5. K. Remmer, *Glastech, Ber.* **25**, 422 (1952).
6. M. L. Hair and I. Altug, *J. Am. Ceram. Soc.* **52**, 65 (1969).
7. A. Kaller, *Silikattechnik* **7**, 380 (1956).
8. E. Brüche and H. Poppa, *Glastech. Ber.* **28**, 232 (1955); E. Brüche and H. Poppa, *ibid.* **29**, 183 (1956); E. Brüche and H. Poppa, *ibid.* **30**, 163 (1957); E. Brüche, *ibid.* **30**, 387 (1957); E. Brüche, K. Peter, and H. Poppa, *ibid.* **31**, 348 (1958); and E. Brüche and K. Peter, *ibid.* **33**, 37 (1960).
9. J. Gotz, *Glastech. Ber.* **40**, 4, 52 (1967).
10. O. Imanaka, *Kikai No Kenkyū (Science of Machinery)* **20**, 60, 357 (1968).
11. D. C. Cornish, *The Mechanism of Glass Polishing, A History and Bibliography* (Taylor & Francis, Philadelphia, 1961).
12. T. Izumitani and S. Harada, *Glass Technol.* **12**, 5, 131 (1971).
13. Y. Wada and H. Hirose, *Kamera Kōgyō Gijutsu Kenkyū Kumiai Shiryo (Camera Industry Technical Res. Ass.)*, *J. Phy. Soc. Japan* **31** (1960).
14. W. B. Hillig, *Advance in Glass Technology* (Plenum Press, New York, 1963), Part 2, pp. 51–52.
15. J. E. Neely and J. O. Mackenzie, *J. Material Science* **3**, 603–609 (1968).
16. F. M. Ernsberger, *J. Am. Ceram. Soc.* **51**, 545–547 (1968).
17. E. W. Taylor, *J. Soc. Glass Technol.* **34**, 69–76T (1950).
18. D. M. Marsh, *Proc. Roy. Soc.* **A279**, 420–435 (1961); and D. M. Marsh, *Proc. Roy. Soc.* **A282**, 33–43 (1964).
19. O. Imanaka, *Denki Shikenjo Iho (Electricity Res. Lab. Rept.)* **25**, 1 (1961).
20. E. Orowan, *Trans. Inst. Eng. Shipbuild Scotl.* **89**, 165 (1945).
21. A. A. Griffith, *Philos. Trans. Roy. Soc.* **A221**, 163 (1920).
22. D. D. Marsh, *Proc. R. Soc. London* **A282**, 33 (1964).
23. R. Hill, *The Mathematical Theory of Plasticity* **97** (Oxford University Press, London and New York, 1950).
24. R. J. Charles and W. B. Hillig, *Symposium sur la rèsistance mèchanique du verre et les moyens de lámèljorer* **USCV** 511–57 (1962).
25. H. Neuber, *Kerbspannungslehre* **142** (Springer-Verlag, Berlin, 1937).
26. T. Izumitani and I. Suzuki, *Glass. Technol. (Great Britian)* **14**, 2, 35–41 (1973), .
27. K. Miyake, *Kamera Kōgyō Gijutsu Kenkyū Kumiai (Camera Industry Technical Res. Ass.)* **19**, 1 (1958).
28. M. Tsunoda, *Senjō Sekkei* **44**, No. 3 (1979).
29. Denshi Kikai Kōgyōkai-hen, *Chō-onpa Ōyō (Electronic Machinery Ass. pub., Ultrasonic Applications)*, Rajikon Gijutsusha (1968).
30. M. Shimakawa, *Chō-onpa Kōgaku—Riron to Jissai (Ultrasonic Engineering—Theory and Practice)* (Kōgyō Chōsakai, Technical Res. Ass., K.K., 1975).
31. T. Fujimoto, *Shin-kaimenkasseizai Nyūmon* (Sanyō Kasei Kōgyō K.K., 1973).
32. Burazā Kōgyō, K.K., Patent Gazette No. 55-28748.
33. Shimada Rika Kōgyō, K.K., Chōfu-shi, Tōkyō-to; Sonikku Ferō K.K., Sagamihara-shi, Kanagawa-ken.
34. F. Tokoroyama and S. Satō, *Kōgaku Gijitsu Konsarutanto* **15**, 5, 26 (1977).
35. P. Benjamin and C. Weaver, *Proc. Roy. Soc.* **A252**, 418 (1959).
36. B. N. Chapman, *J. Vac. Sci. Technol.* **11**, 106 (1974).
37. A. Kinbara and H. Fujiwara, *Hakumaku (Ōyō Butsuri Gakukai Sensho 3)*, (Shōkabō, 1980), p. 125.
38. C. Kanamori and T. Izumitani, *Shōwa 56-nen Shunki Dai-28-kai Ōbutsu Rengō Kōen-kai Yokōshū (1981—Spring, 28th Applied Physics Conf. Proceedings)*, p. 291.
39. Y. Nakajima, C. Kanamori, and T. Izumitani, *Shōwa 56-nen Shunki Dai-28-kai Ōbutsu Rengō Koen-kai Yokōshū (1981—Spring, 28th Applied Physics Conf. Proceedings)*, p. 290.

40. O. S. Heavens, *J. Phys. Rad.* **11**, 355 (1950).

41. P. Benjamin and C. Weaver, *Proc. Roy. Soc.* **A254**, 163 (1960).

42. S. Adachi, E. Miyade, and T. Izumitani, *J. Non-Crystal. Sol.* **42**, 569 (1980).

43. K. Kinoshita, *Progress in Optics. IV*, (North Holland Publishing Company, Amsterdam, 1965), p. 115.

44. H. Schroeder, *Ann. Physik* **39**, 55 (1941).

45. Y. Asahara and T. Izumitani, *J. Non-Crystal. Sol.* **42**, 269 (1980).

46. C. Kanamori and T. Izumitani, Hoya Seminar, (Cologne, 1982).

47. Y. Nakajima and T. Izumitani, *XIIth International Glass Congress*, poster session (Albuquerque, NM, 1980).

CHAPTER 5

New Glasses

5.1. Optical Glasses

5.1.1. Progress in Optical Glasses

As mentioned earlier, the introduction of rare elements such as La, Th, Ta, etc., into borate glasses greatly expanded the range of optical glasses. Subsequent improvements in optical glass compositions have moved in the following directions:

- High-refraction, low-dispersion glasses containing no Cd or Th.
- High-dispersion glasses with improved hardness.
- Low-dispersion fluorophosphate glasses.
- Low-specific-gravity, high-refraction glasses.
- Improvement of chemical durability.
- Improvement of coloring.
- Glass compositions containing less of the high-cost raw material Ta_2O_5.

5.1.1.1. High-Refraction, Low-Dispersion Glasses

One conspicuous fact is that CdO and ThO_2 offer a very broad devitrification region and great stability against devitrification as divalent and tetravalent components in high-refraction, low-dispersion glasses. Because of pollution problems, however, the use of these components has been prohibited. Accordingly, $(Gd_2O_3 + Y_2O_3)$ and Yb_2O_3 are now being used instead of ThO_2, and $(ZnO + Nb_2O_5)$ and $(BaO + Nb_2O_5)$ are being used instead of CdO (Refs. 1–4).

Furthermore, efforts have been made to raise the index of refraction to a value above 1.85 in borate glasses.[1-4] In addition, Bremer of Leitz found that greater amounts of rare elements such as La_2O_3, etc., can be introduced in borosilicate glasses than in borate glasses resulting in a higher index of refraction.[1-4] However, no glass has been found that can be consistently supplied.

5.1.1.2. High-Dispersion Glasses

In the past, compositions containing large amounts of PbO were used to produce high-dispersion glasses. However, such glasses had a low hardness and insufficient chemical durability. Accordingly, SiO_2–TiO_2–Nb_2O_5–RO–Li_2O glasses were developed to replace SiO_2–PbO–R_2O glasses.[1-3] For example, whereas the Knoop hardness of SF 6 was 365 kg/mm², the new glass system offers a high hardness and a good chemical durability at a low specific gravity. The hardness has been increased to as high as 550 kg/mm². However, coloring is stronger than in PbO systems, and the stability against devitrification is insufficient. Furthermore, phosphate glasses containing TiO_2 or Nb_2O_5 compositions and SiO_2–P_2O_5 glass have been investigated as high-glasses.[1-4] These glasses are still impractical because they have a low hardness and problems in melting.

5.1.1.3. Low-Dispersion Glasses

The P_2O_5–AlF_3–RF_2–RF fluorophosphate glasses have been developed.[1-6] For example, FCD 10 is a low-dispersion glass with an extremely large Abbe number in which $n_d = 1.457$ and $v_d = 90.8$. As described later, this glass shows great abnormal dispersion characteristics in the short-wavelength region and at the same time is an athermal glass in which the temperature coefficient of the light path length is extremely small. This glass is extremely useful for telescopic achromatic lenses.

5.1.1.4. Low-Specific-Gravity, High-Refraction Glasses

The glass BaFD 15 is a borosilicate glass that contains Li_2O, TiO_2, and Nb_2O_5. At $n_d = 1.70$, it shows approximately the same index of refraction as BaFD 7. However, the respective specific gravities are 2.99 and 3.47; thus, BaFD 15 is lighter and can be used advantageously in eyeglasses.

5.1.1.5. Chemical Durability Improvement

The basic solution to the problems of dimming, staining, and latent scratching lies in the improvement of the chemical durability of optical glasses. The chemical durability has been improved in many glasses. For example, BaCD 16, which has long been given as an example of a glass that is especially susceptible to dimming and staining, has been improved in terms of chemical durability by the addition of La_2O_3 and ZrO_2. The results of this improvement are shown in Table 5.1.

5.1.1.6.   Improvement of Coloring

The coloring of optical glasses is influenced by the glass composition, raw materials, amounts of impurities in the refractory, amount of cullet used, melting temperature, melting atmosphere, and dissolution of Pt from Pt pots. Coloring has been improved in 32 types of glass (Fig. 5.1a).

5.1.1.7.   Decrease in the Amounts of Expensive Raw Materials

In recent years, there has been a conspicuous increase in the cost of already expensive raw materials, especially Ta_2O_5. As this occurred, glasses were developed that contained decreased amounts of Ta_2O_5 or contained no Ta_2O_5.[1-3] This was accomplished by replacing Ta_2O_5 with La_2O_3, (Y_2O_3, Gd_2O_3), and Nb_2O_5. In Figure 5.1, types of optical glasses in which the coloring, durability, and hardness have been conspicuously improved are shown in n_d–v_d graphs.

5.1.2.   Abnormal Dispersion Glasses

Figure 5.2 shows the behavior of some optical glasses where the relationship between index of refraction and wavelength differs from that of ordinary optical glasses at short wavelengths and again at long wavelengths.[7-9] In ordinary optical glasses, there is a roughly linear relationship between the partial dispersion ratio $P_{i,g}$ [$= (n_i - n_g)/(n_F - n_C)$] that expresses the partial dispersion characteristics in the short-wavelength range and Abbe number v_d, and also between the partial dispersion ratio $P_{C,t}$ [$= (n_C - n_t)/(n_F - n_C)$] that ex-presses the partial dispersion characteristics in the long-wavelength region and Abbe number (Fig. 5.3). Glasses that depart conspicuously from this linear relationship are called abnormal-partial-dispersion glasses. As seen in Fig. 5.4, the use of abnormal dispersion glasses to construct a lens is extremely effective in correcting the chromatic aberration of the lens. The glasses FCD 10, ADC 1, ADC 2, ADF 4, etc., are representative optical glasses that display abnormal partial dispersion. In this chapter, we will discuss what abnormal dispersion glasses really are, as well as the causes of abnormal dispersion.

According to Lorentz's theoretical equation, the index of refraction n can be expressed as a function of the wavelength λ in a wavelength region where there is no absorption:

$$n^2 - 1 = \sum_i \frac{KN_i f_i}{\left(1/\lambda_i^2 - 1/\lambda^2\right)} . \tag{5.1}$$

Here, λ_i is the wavelength of inherent absorption, f_i is the oscillator strength of inherent absorption, N_i is the number of ions per unit volume, and K is a constant. From this equation, we may say that the dispersion is determined by several inherent absorptions. In the case of optical glasses, there is absorption by bridging oxygen ions, absorption by nonbridging oxygen ions, and (when the glass contains Pb ions) absorption by Pb ions in the ultraviolet region, and there is also absorption in the infrared region. Assuming that these absorptions determine the dispersion, then Lorentz's theoretical equation can be rewritten as follows:

$$n^2 - 1 = \frac{KN_1 f_1}{1/\lambda_1^2 - 1/\lambda^2} + \frac{KN_2 f_2}{1/\lambda_2^2 - 1/\lambda^2}$$
$$+ \frac{KN_3 f_3}{1/\lambda_3^2 - 1/\lambda^2} - \frac{KN_4 f_4}{1/\lambda^2 - 1/\lambda_4^2} . \tag{5.2}$$

Table 5.1   Examples of chemical durability improvement.

		D_w (10^{-3} mg/cm^2·h)	D_0 (10^{-3} mg/cm^2·h)	D_{NaOH} (mg/cm^2·5 h)	
BaCD 16	Old	0.25	57.4	0.55	
	New	0.22	11.1	0.11	
				D_{STPP} (mg/cm^2·h)	T_{Blue} (h)
LaFL 3	Old	0.19	144	0.5	0.8
	New	0.03	4.6	0.01	45

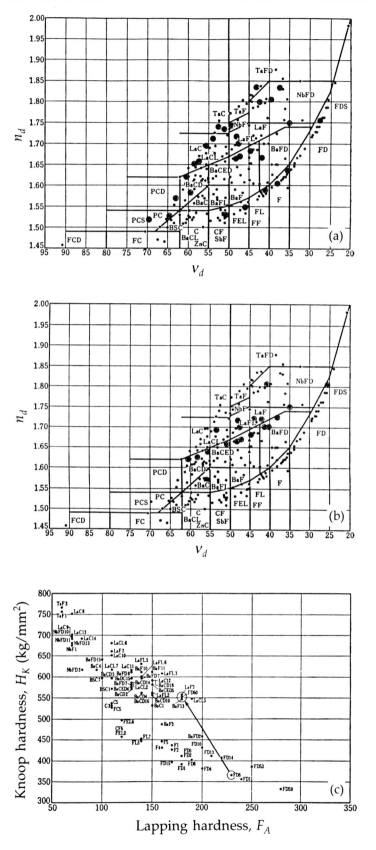

Figure 5.1.   Optical glasses with conspicuously improved characteristics: (a) improvement of coloring, (b) improvement of durability, and (c) improvement of hardness.

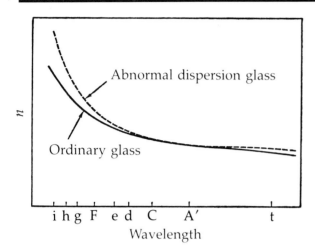

Figure 5.2. Abnormal dispersion characteristics seen in the wavelength dependence of the index of refraction.

The first term in Eq. (5.2) arises from absorption by bridging oxygen ions, λ_1 is the peak wavelength of this absorption, and $KN_1 f_1$ is the intensity. Similarly, the second term arises from absorption by nonbridging oxygen ions, and the third term arises from absorption by lead ions (this third term does not occur in the case of glasses that do not contain lead ions). The fourth term arises from infrared absorption. Here, the wavelengths and intensities of these inherent absorptions are determined by substituting the index of refraction from the t line to the i line into this equation, and an attempt is made to clarify the causes of the abnormal dispersion characteristics shown by abnormal dispersion glasses. This clarification is based on measurements of the vacuum ultraviolet-reflection spectrum and infrared absorption spectrum.

Figure 5.5a and b shows the variation of $P_{i,g}$ and v and the variation of $P_{C,t}$ and v that are observed when λ_i and $KN_i f_i$ are varied with K 7 as a starting point. Here, because absorption is known to occur in the extreme ultraviolet region (around 20 eV) in silicate glasses,[10] this wavelength λ_0 was assumed to be 0.0684 μm[11]; actual measured values were used for the remaining λ_i. As shown in Fig. 5.5, v decreases, $P_{i,g}$ increases, and $P_{C,t}$ decreases as $\lambda_{1,2,3}$ moves toward longer wavelengths, and as $KN_i f_i$ increases. However, as λ_0 and $KN_0 f_0$ increase, v increases (the reverse of what occurs when $\lambda_i = 1,2,3$) and $P_{i,g}$ either shows almost no change, or shows a moderate increase. The value $P_{C,t}$ shows a moderate decrease. As seen in Fig. 5.6b, when ultraviolet absorption

occurs at shorter wavelengths, $P_{i,g}$ is unaffected and shows almost no change. As the absorption moves toward longer wavelengths, however, $P_{i,g}$ is affected, and shows an increase. As seen in Fig. 5.6a, on the other hand, v decreases as the absorption moves toward longer wavelengths. Accordingly, $P_{i,g}$ and v varies as shown in Fig. 5.7 with a decrease in λ. The effect of an increase in $KN_i f_i$ on $P_{i,g}$, $P_{C,t}$, and v is the same as the effect of a shift of λ_i toward longer wavelengths. The infrared absorption wavelength λ_4 does not affect v; an increase in $KN_4 f_4$ causes a slight decrease in v, a decrease in $P_{i,g}$, and an increase in $P_{C,t}$.

5.1.2.1. Causes of Abnormal Characteristics of Fluorophosphate Glass FCD 10

Table 5.2 shows calculated values of $KN_i f_i$ and λ_i (except that $\lambda_i = 1,2,3$ are measured values and λ_4 values are mean values) for the normal dispersion glasses K 7 and F 2, and for the abnormal dispersion glass FCD 10. Compared with K 7, F 2 shows an increase in $P_{i,g}$ and a decrease in the Abbe number because of absorption by Pb^{2+} ions at 207 and 251 nm (see Fig. 5.5 and Table 5.2). The line connecting the $P_{i,g}$, v_d coordinates of K 7 and F 2 represents normal dispersion glasses.

As shown in Table 5.2, FCD 10 has the following special features: the wavelength λ_1 of ultraviolet absorption by $(AlF_6)^{3-}$ is shorter (0.110 μm) than the wavelengths of absorption by Si–O bridging oxygens (0.122 μm). There is no absorption ($KN_2 f_2$) by nonbridging oxygens, and the intensity $KN_4 f_4$ of infrared absorption by P_2O_5 is low. Figure 5.5a shows that these factors cause a departure from the normal curve. One cause of the abnormal partial dispersion characteristics of FCD 10 is that absorption occurs at shorter wavelengths. This was also confirmed by the vacuum ultraviolet reflection spectrum measurements in Fig. 5.8a. Another cause of the abnormal dispersion characteristics of FCD 10 is that the intensity $KN_4 f_4$ of infrared absorption is low, as anticipated from Table 5.2. In this case as well, the low infrared absorption of the fluorophosphate glass FCD 10 in the vicinity of 9 μm was confirmed by the absorption spectrum in the infrared region (as shown in Fig. 5.8b). In short, the abnormal dispersion characteristics of FCD 10 are attributable to the following facts: because FCD 10 is a phosphate glass with a high fluoride content, ultraviolet absorption occurs at lower wavelengths, and infrared absorption is small.

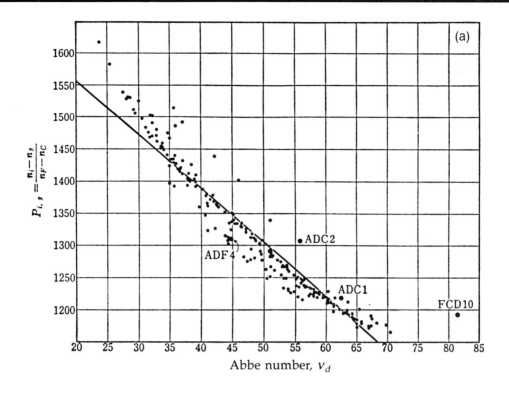

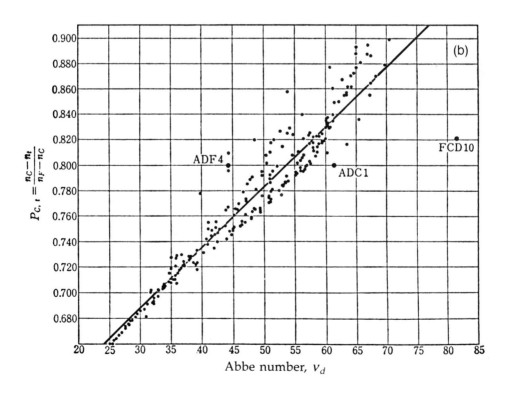

Figure 5.3. Relationship between partial dispersion and the Abbe number: (a) short wavelengths and (b) long wavelengths.

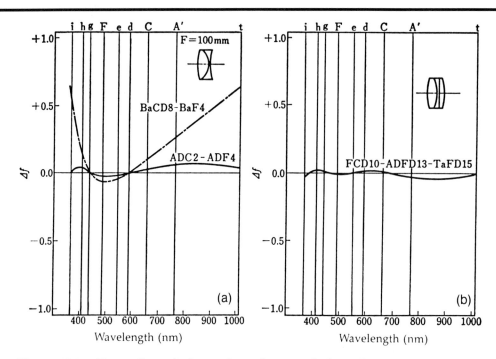

Figure 5.4. Examples of aberration characteristics of super-achromatic lenses: (a) ADC 2–ADF 4 and (b) FCD 10–ADF 8–TaFD 15.

Table 5.2.   Wavelengths and intensities of ultraviolet and infrared inherent absorptions (calculated).

	FCD 10		K 7		F 2	
n_d	1.45650		1.51112		1.62004	
v_d	90.77		60.41		36.37	
$P_{i,g}$	1.1725		1.2179		1.4194	
$P_{C,t}$	0.8357		0.8315		0.7175	
λ_0 (μm)	0.0679		0.0684		0.0684	
$KN_0 f_0$	189.13		151.40		158.62	
λ_1 (μm)	0.110	0.122	0.106	0.122	0.106	0.122
$KN_1 f_1$	18.02	0.85	19.10	18.27	17.38	14.89
λ_2 (μm)	—		0.172		0.181	
$KN_2 f_2$	—		1.93		9.94	
λ_3 (μm)	—		—		0.207	0.251
$KN_3 f_3$	—		—		1.04	0.40
λ_4 (μm)	12.0		9.8		9.8	
$KN_4 f_4$	0.00453		0.00825		0.00840	
calc-meas						
Δn_t	– 0.00000		– 0.00000		0.00000	
Δn_s	– 0.00001		– 0.00000		– 0.00000	
Δn_r	– 0.00001		– 0.00000		– 0.00000	
Δn_C	– 0.00000		– 0.00000		– 0.00000	
$\Delta n_{C'}$	– 0.00000		– 0.00000		– 0.00000	
Δn_D	0.00000		– 0.00000		– 0.00000	
Δn_d	0.00000		– 0.00000		– 0.00000	
Δn_e	0.00000		– 0.00000		– 0.00000	
Δn_F	– 0.00000		– 0.00000		– 0.00000	
$\Delta n_{F'}$	– 0.00000		– 0.00000		– 0.00000	
Δn_l	– 0.00001		– 0.00000		0.00000	
Δn_h	– 0.00001		0.00000		0.00001	
Δn_i	– 0.00001		– 0.00000		– 0.00000	

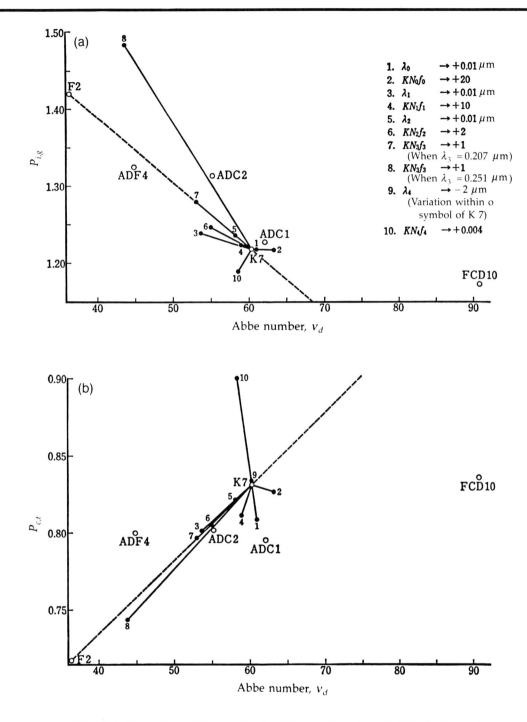

Figure 5.5. **(a) Variation of $P_{i,g}$ and vd with varying λ_i and $KN_i f_i$; (b) variation of $P_{C,t}$ and vd with varying λ_i and $KN_i f_i$.**

5.1.2.2. Abnormal Dispersion Characteristics of Phosphate Glasses ADC 1 and 2

Table 5.3 shows λ_i and $KN_i f_i$ for ADC 1 and BK 6 with the same Abbe number. In addition, the vacuum ultraviolet reflection spectra and infrared absorption spectra are shown in Fig. 5.9a and b.

Compared with BK 6, ADC 1 shows higher $KN_0 f_0$ and $KN_1 f_1$, a lower λ_2, and a lower $KN_4 f_4$. The fact that ADC 1 shows a higher $P_{i,g}$ apparently results from the fact that the infrared absorption ($KN_4 f_4$) is small and $KN_1 f_1$ is high (see Fig. 5.5a). It appears from this that the abnormal characteristics of ADC 1 result from the fact that the absorption by

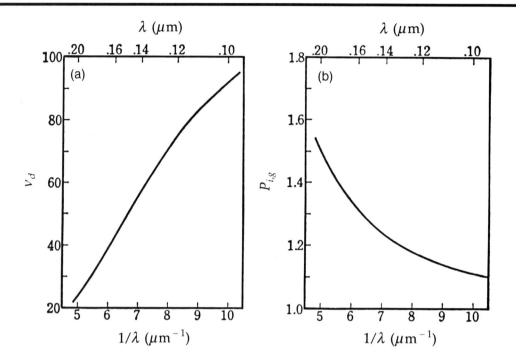

Figure 5.6. (a) Relationship between ultraviolet absorption wavelength λ and ν_d; (b) relationship between ultraviolet absorption wavelength λ and $P_{i,g}$.

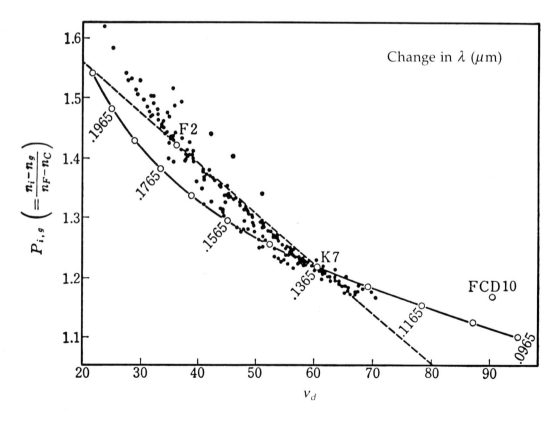

Figure 5.7. Variation of $P_{i,g}$ and νd with varying ultraviolet absorption wavelength λ.

P_2O_5 in the infrared region is small (see Fig. 5.9b), and the fact that there is strong absorption by P_2O_5 in the ultraviolet region (as shown by Fig. 5.9a).

In the case of ADC 2, Ti^{4+} is added to produce absorption at the longer wavelengths of the ultraviolet region. Accordingly, we may take the view that in addition to the large ultraviolet absorption by P_2O_5, a large absorption at λ_3 (in the vicinity of 234 nm) by Ti^{4+} contributes to $P_{i,g}$ so that ADC 2 shows a higher partial dispersion ratio $P_{i,g}$ than K 10 with the same Abbe number. Furthermore, the high Abbe number of ADC glasses results from the fact that λ_2 is low, i.e., the fact that absorption by nonbridging oxygens in P_2O_5 occurs at shorter wavelengths.

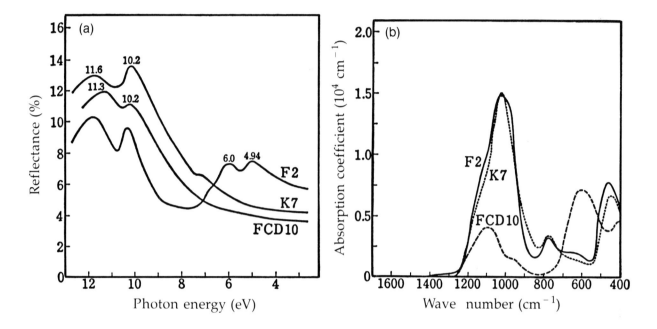

Figure 5.8.   (a) Reflection spectra of FCD 10, K 7, and F 2 (vacuum ultraviolet to visible region) and (b) infrared absorption spectra of FCD 10, K 7, and F 2.

Table 5.3.   Characteristics of abnormal partial dispersion glasses compared with those of ordinary glasses.

	λ_0	KN_0f_0	λ_1	KN_1f_1	λ_2	KN_2f_2	λ_3	KN_3f_3	λ_4	KN_4f_4	n_d	ν_d	$P_{i,g}$	$P_{C,t}$
ADC 1	0.0634	225.13	0.127	40.85	0.155	0.790	—	—	10.0	0.00793	1.62000	62.19	1.2268	0.7955
BK 6	0.0684	166.29	0.108	18.20	0.175	0.509	—	—	9.7	0.00934	1.53113	62.15	1.2019	0.8500
			0.124	14.09										
			0.141	4.52										
ADC 2	0.0634	228.32	0.127	32.10	0.155	0.660	0.234	0.687	10.0	0.01013	1.59700	55.29	1.3126	0.8022
			0.139	1.00										
K 10	0.0684	143.09	0.106	18.83	0.172	2.095	0.196	0.140	9.8	0.00832	1.50137	56.41	1.2463	0.8213
			0.122	18.00			0.237	0.088						
ADF 4	0.0684	201.07	0.111	3.01	0.172	3.188	0.180	1.899	7.7	0.01298	1.61250	44.87	1.3237	0.8002
			0.139	20.02			0.238	0.468						
LLF 4	0.0684	150.70	0.106	18.71	0.175	5.883	0.199	0.389	9.8	0.00836	1.56138	45.23	1.3398	0.7608
			0.122	17.50			0.242	0.304						

5.1.2.3. Abnormal Dispersion Characteristics of the Kurz Flint Glass ADF 4

Table 5.3 shows that ADF 4 is a borate glass containing Pb. Compared with the silicate glass LLF 4 with the same Abbe number, ADF 4 shows a higher intensity of infrared absorption (KN_4f_4) by B_2O_3 (see Fig. 5.10b). It appears that this term contributes to the dispersion so that the partial dispersion ratio $P_{C,t}$ is increased. Furthermore, the fact that ultraviolet absorption (KN_1f_1) by B_2O_3 is small also contributes to this (see Fig. 5.10a).

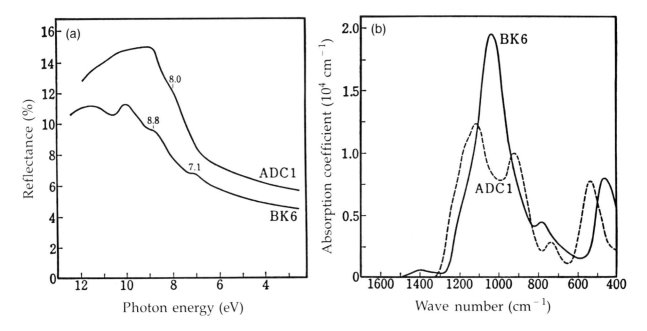

Figure 5.9.   (a) Reflection spectra of ADC 1 and BK 6 (vacuum ultraviolet to visible region) and (b) infrared absorption spectra of ADC 1 and BK 6.

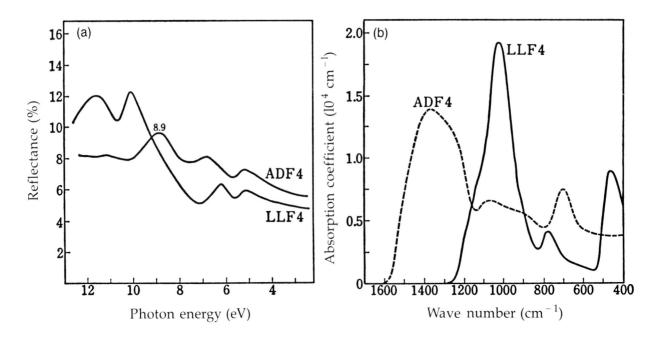

Figure 5.10.   (a) Reflection spectra of ADF 4 and LLF 4 (vacuum ultraviolet to visible region) and (b) infrared absorption spectra of ADF 4 and LLF 4.

5.1.3. Athermal Glasses

An athermal glass is a glass in which the temperature coefficient of the light path length is zero. Glasses marked with a ⊙ in the $n_d - v_d$ graph (Fig. 5.11) show athermal properties. In an optical system that follows temperature changes, the focal length of the lens made of athermal glasses does not vary with temperature. Accordingly, athermal glasses are very useful for laser oscillators and telescopes that are sent up in artificial satellites, etc.

When the temperature changes, the light path length changes as a result of thermal expansion and temperature variation of the index of refraction. The temperature variation of the light path length dS/dT is given by the following equation:

$$\frac{dS}{dT} = (n - 1)\,\alpha + \frac{dn}{dT} \quad . \tag{5.3}$$

Here, n is the index of refraction and α is the coefficient of linear expansion. For dS/dT to equal zero, dn/dT must be less than zero.

The temperature coefficient of the index-of-refraction dn/dT is derived by differentiating the Lorentz-Lorenz equation and is given by the following equation:

$$\frac{dn}{dT} = A\,(\varphi - \beta) \quad . \tag{5.4}$$

Here,

$$A = \frac{(n^2 - 1)(n^2 + 2)}{6n} \quad . \tag{5.5}$$

$$\varphi = \frac{d \ln \alpha_e}{dT} \quad . \tag{5.6}$$

$$\beta = \frac{d \ln V}{dT} \doteq 3\,\alpha \quad . \tag{5.7}$$

Here, α_e is the electron polarizability, φ is the temperature coefficient of the polarizability, V is the molecular volume, and β is the coefficient of volume expansion. Figure 5.12 shows the relationship between φ and α. A glass in which dn/dT is

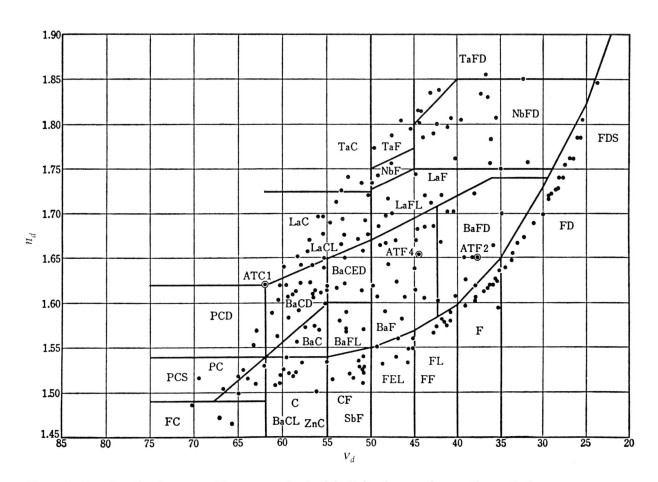

Figure 5.11.   Graph of $n_d - v_d$. Glasses marked with ⊙ in the graph are athermal glasses.

positive, zero, or negative is obtained depending on whether φ is greater than, equal to, or less than 3α. Regarding φ, then, assuming that a change in temperature causes a change in the interionic distance and that the polarizability is increased by an increase in the interionic distance, we may write the following equation[12,13]:

$$\varphi = \left(\frac{\partial \ln \alpha_e}{\partial \ln r}\right)_T \left(\frac{d \ln r}{dT}\right) = \gamma \cdot \alpha \quad . \qquad (5.8)$$

Here, r is the interionic distance, γ is the change in polarizability that accompanies a change in the interionic distance, and α is the temperature variation of the interionic distance, i.e., the coefficient of linear expansion. From Eq. (5.4), we can write

$$\frac{dn}{dT} = \frac{(n^2 - 1)(n^2 + 2)}{6n}(\gamma - 3)\alpha \quad . \qquad (5.9)$$

Accordingly, dn/dT is positive when $\gamma > 3$, negative when $\gamma < 3$, and zero when $\gamma = 3$. Then, from Eq. (5.3),

$$\begin{aligned}
\frac{dS}{dT} &= (n - 1)\alpha + \frac{dn}{dT} \\
&= (n - 1)\alpha + \frac{(n^2 - 1)(n^2 + 2)}{6n}(\gamma - 3)\alpha \\
&= \alpha\frac{(n^2 - 1)(n^2 + 2)}{6n} \\
&\quad \left[\frac{6n}{(n + 1)(n^2 + 2)} + (\gamma - 3)\right] \quad . \quad (5.10)
\end{aligned}$$

Because α and $(n^2 - 1)(n^2 + 2)/6n$ are positive, $[6n/(n + 1)(n^2 + 2)] + (\gamma - 3)$ must equal zero for dS/dT to equal zero. Accordingly,

$$\gamma = 3 - \frac{6n}{(n + 1)(n^2 + 2)} \quad . \qquad (5.11)$$

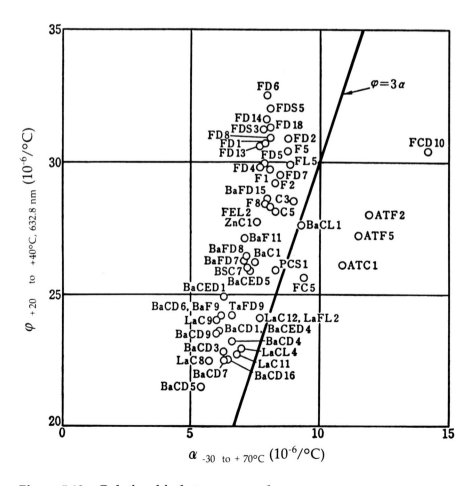

Figure 5.12.   Relationship between φ and α.

Figure 5.13 shows a plotting of the relationship between γ and $g(n) \equiv 6n/(n + 1)(n^2 + 2)$. Glasses on the straight line in the figure are athermal glasses. Glasses in which the value of γ is 2.1 to 2.3 with n = 1.45 to 1.90 are athermal glasses. In ordinary optical glasses, γ is usually large so that when the temperature rises, there is a conspicuous change in polarizability and increase in index of refraction accompanying the increase in the interatomic distance; as a result, the temperature coefficient of the light path length is positive. But in the case of athermal glasses, γ is small, and the change in thermal expansion is greater than the change in polarizability accompanying an increase in the interatomic distance caused by a rise in the temperature. The index of refraction decreases with the temperature and compensates for a positive effect on the light-path length, which is caused by thermal expansion, so that the temperature variation of the optical-path length is zero. In other words, ordinary glasses are glasses in which a rise in temperature causes an increase in electron polarizability, whereas athermal glasses may be characterized as glasses in which the interatomic distance shows a conspicuous increase with little change in polarizability.

5.1.3.1.   Relationship of φ and γ to Glass Cationic Field Strength

In regard to φ_i, additivity of $\Sigma\varphi_i x_i$ holds true for the glass components. The additive factors φ_i of various component ions are plotted against the respective cationic field strengths z/a^2 in Fig. 5.14. The factor φ_i decreases with an increase in cationic field strength; as seen from the fact that $\varphi_i = \gamma_i\alpha_i$, the temperature coefficient of polarizability φ_i is governed primarily by the increase in the interatomic distance. On the other hand, as shown by Fig. 5.15, γ_i increases with an increase in z/a^2, indicating that the change in oxygen–ion polarizability accompanying an increase in the interatomic distance is greater in the case of ions with a large polarizing power than in the case of ions with a small polarizing power. In regard to the small γ_i value observed for the B^{3+} ion, it is surmised that intermolecular bonds of triangular coordination are actually present so that the actual expansion coefficient for the B^{3+} ion is greater than the value of α_i determined by ze/a^2. It is thought that for this reason the apparent value of γ_i is small, but that γ_i actually has a larger value relative to the value of z/a^2 for B^{3+} (as in the case of Si).

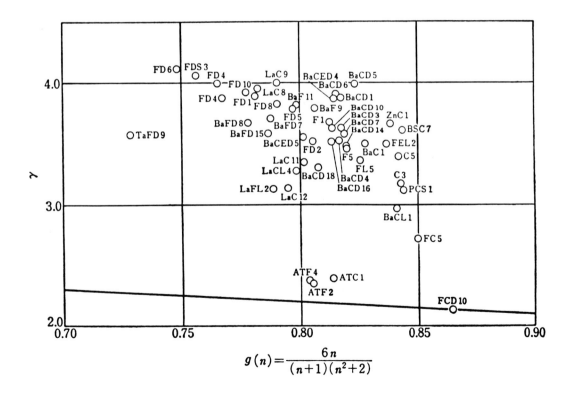

Figure 5.13.   Relationship between γ and $g(n)$.

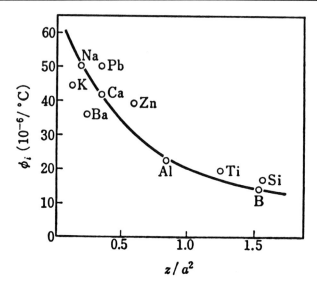

Figure 5.14. Relationship between cationic field strength (z/a^2) and temperature coefficient of electron polarizability (ϕ_i) of ions.

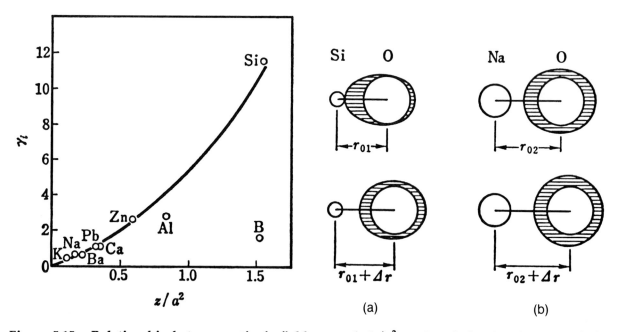

Figure 5.15. Relationship between cationic field strength (z/a^2) and variation in electron polarizability caused by temperature variation of interatomic distance (γ_i).

5.2. Laser Glasses

Laser light consists of in-phase light of more or less constant wavelength. A special feature of laser light is that it possesses coherency, unlike ordinary light. In 1962, Snitzer discovered Nd laser glass. However, the use of this glass ceased around 1972 with ruby lasers and yttrium-aluminum-garnet (YAG) lasers taking over the mainstream of solid lasers. Currently, glass lasers are again being used in the high-power laser devices employed in laser fusion.

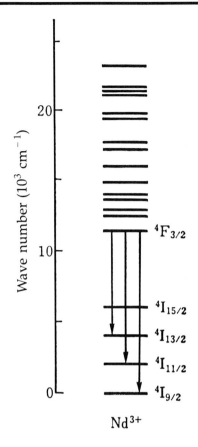

Figure 5.16.   Energy levels of Nd^{3+} ion.

An Nd glass laser contains Nd^{3+} ions, which have the energy levels shown in Fig. 5.16, as activated ions. Thus, this laser uses the properties of Nd^{3+} ions as four-level laser ions. Electrons excited to higher energy levels from the ground level of Nd^{3+} move to the $^4F_{3/2}$ level by nonradiative transitions so that a population inversion is formed between the $^4F_{3/2}$ level and ^4I levels; the electrons then move from the $^4F_{3/2}$ level to the ^4I levels by radiative transitions. Among the four ^4I levels, the radiative transition from $^4F_{3/2}$ to $^4I_{11/2}$ is the strongest; this transition is used as 1.06-μm laser light. However, nonradiative transitions also occur between the $^4F_{3/2}$ level and ^4I levels. Therefore, to achieve efficient laser light oscillation, it is important to suppress nonradiative transitions as far as possible so that the energy transition between the levels consists of radiative transitions.

In 1978, a silicate laser glass (LSG 91H) was used in the Shiva system at Lawrence Livermore National Laboratory; at the same time, an edge-coated glass was developed for use on the laser disk, and a Faraday rotation glass was developed for use as an optical isolator for high-output laser light. A phosphate laser glass (LHG 8) that showed high cross section of stimulated emission and had a small nonlinear coefficient was used in the Omega system built a little later at the University of Rochester. A fluorophosphate glass LHG 10 with an extremely small nonlinear coefficient was developed for use in the Nova device under construction in 1982 at Lawrence Livermore National Laboratory. However, because the laser damage threshold was not high enough, the use of this glass was abandoned, and the phosphate laser glass LHG 8 is now being used instead. Recently, an antireflection coating with a high damage threshold and a frequency conversion filter for the purpose of obtaining short-wavelength laser light have also been developed. In the future, laser glasses with a high-gain saturation value and highly doped laser glasses with a long lifetime will probably also be developed. Thus, laser glasses have been developed as oscillator and amplifier materials for use in high-power laser fusion devices. However, because high repetition rates have become possible as a result of the development of athermal glasses, we may expect to see laser glasses used in laser materials for more popular applications.

5.2.1.   Necessary Characteristics of Laser Glasses

5.2.1.1.   Laser Oscillation Conditions and Gain Coefficient

The initial conditions of laser oscillation are given by the following equation:

$$R_1 R_2 \exp\left[(g_0 - \gamma)\, 2l\right] = 1 \quad . \tag{5.12}$$

Here, R_1 and R_2 are the reflectances of the mirrors, g_0 is the gain coefficient, γ is the loss coefficient, and l is the length of the laser medium. Oscillation begins when the gain exceeds the loss because of reflection, absorption, and scattering, etc., in the medium. For this to occur, however, it is necessary that the number of atoms at the higher energy level be greater than the number of atoms at the lower energy level (population inversion: $\Delta N = N_2 - N_1 > 0$). Here, the gain coefficient is given by the following equation:

$$g_0 = \Delta N \cdot \sigma \quad . \tag{5.13}$$

Where ΔN is the population inversion and σ is the cross section of stimulated emission.

5.2.1.2. Stimulated-Emission Cross Section and Population Inversion

The cross section of stimulated emission σ is a characteristic value that is determined by the composition of the medium, i.e., the glass composition. This value can be determined semi-empirically (as shown below) using Judd-Ofelt's theory concerning electric dipole transitions of rare-earth ions in (or around) a crystalline field.[14] Izumitani et al.[15] also determined $\sigma_{0.88}$ for

$$^4F_{3/2} \rightarrow {}^4I_{9/2}$$

transition from the absorption at 0.88 μm using Fuchtbauer-Ladenburg's equation, and also determined $\sigma_{1.06}$ from this and the emission intensity ratio of 0.88 to 1.06 μm.

$$S_{JJ'} = \sum_{2,4,6} \Omega_t |\langle S'L'J' \| \mathbf{U}^{(t)} \| SLJ \rangle|^2 \quad . \tag{5.14}$$

Here, Ω_t is an intensity parameter, and $\langle \| \mathbf{U}^{(t)} \| \rangle$ is a unit tensor operator. The line intensity $S_{JJ'}$ shows the following relationship with the integrated absorption intensity $\int k d\lambda$:

$$\int k d\lambda = \frac{8\pi^3 e^2 \lambda N_0}{3nch(2J + 1)} \frac{(n^2 + 2)^2}{9} S_{JJ'} \quad . \tag{5.15}$$

Where, λ is the wavelength, N_0 is the ion concentration per unit volume, J is the total angular momentum quantum number, and n is the index of refraction. If Ω_t is determined by the method of least squares for Eqs. (5.14) and (5.15), and the line intensity of radiation $S_{JJ'}$ is determined using this, then the radiative transition probability A can be determined from the following equation:

$$A_{JJ'} = \frac{64\pi^4 e^2 n}{3h (2J + 1) \lambda^3} \frac{(n^2 + 2)}{9} S_{JJ'} \quad . \tag{5.16}$$

The total radiative transition probability is the sum of the probabilities of transitions from the $^4F_{3/2}$ level to all final levels, and is given by the following equation:

$$A_{\text{rad}} = \sum_{J'} A_{JJ'} \quad . \tag{5.17}$$

The coefficient of stimulated emission is given in the form of $A_{JJ'}$ divided by the half-width of the fluorescence spectrum $\Delta\lambda$:

$$\sigma = \frac{\lambda^4}{8\pi ch^3} \frac{A_{JJ'}}{\Delta\lambda}$$

$$= \frac{8\pi^3 e^2}{3ch(2J + 1)} \frac{1}{\Delta\lambda} \frac{(n^2 + 2)}{9n} S_{JJ'} \quad . \tag{5.18}$$

Accordingly, to obtain a large σ, the spectral width must be narrow, the index of refraction must be high, and the line intensity must be great, i.e., the medium must be a medium with a large intensity parameter Ω_t. We will discuss what determines Ω_t in a later section. In addition, the population inversion ΔN is proportional to the fluorescence lifetime τ and the pumping efficiency p, and can be expressed as follows (where N_0 is the number of atoms in the ground state):

$$\Delta N \propto N_0 p \cdot \tau \quad . \tag{5.19}$$

In other words, ΔN is determined by the Nd^{3+} concentration and the fluorescence lifetime. The fluorescence lifetime is given by the radiative transition probability A_{rad} and the nonradiative transition probability W_{nr} in the following relationship:

$$\tau = \frac{1}{(A_{\text{rad}} + W_{\text{nr}})} \quad . \tag{5.20}$$

Accordingly, to increase ΔN while maintaining a high σ, it is necessary to minimize nonradiative transitions.

The quantum efficiency η is defined by the following equation:

$$\eta = \frac{A_{\text{rad}}}{A_{\text{rad}} + W_{\text{nr}}} = A_{\text{rad}} \cdot \tau \quad . \tag{5.21}$$

Furthermore, the gain coefficient is given by the following equation:

$$g_0 = \sigma \cdot \Delta N \propto \frac{A_{\text{rad}}}{\Delta\lambda} \cdot \Delta N = \frac{\int I d\lambda}{\Delta\lambda} = I_P \quad . \tag{5.22}$$

Accordingly,

$$g_0 \propto \frac{A_{\text{rad}}}{\Delta\lambda} \cdot \tau = \frac{\eta}{\Delta\lambda} \quad . \tag{5.23}$$

Here, I is the intensity of the emission spectrum, and

$$\int I d\lambda = A_{\text{rad}} \cdot \Delta N \quad .$$

From Eq. (5.23), we can say that the gain coefficient is proportional to the quantum efficiency. Furthermore, from Eq. (5.22), g_0 can be empirically estimated according to the peak intensity of the fluorescence spectrum.[15] In laser glasses, a high coefficient of stimulated emission σ and a long mean lifetime of fluorescence τ are generally desirable.

5.2.1.3. Gain Saturation and Energy Extraction

When the intensity of incident light is low, a high σ value leads to an increase in the gain. When the incident light intensity is high, however, a saturation of gain occurs as shown in Fig. 5.17 (Ref. 16). The output is given by Franz-Nodvik's equation:

$$\Phi_{\text{out}} = \Phi_s \ln\left[1 + g\left(\exp\frac{\Phi_i}{\Phi_s} - 1\right)\right] \quad . \tag{5.24}$$

Here, g is the small-signal gain coefficient, Φ_i is the input intensity, and Φ_s is the saturation intensity. The variable Φ_s can be theoretically expressed as follows[16]:

$$\Phi_s = \frac{h\nu}{\gamma\sigma} \quad , \tag{5.25}$$

where $\gamma = 1$ to 2.

The value of γ depends on the relaxation time of the ${}^4I_{11/2}$ state; when this relaxation time is long, γ equals 1. Consequently, gain saturation is less likely to occur in the case of a small σ value. As shown in Fig. 5.18, measured values of the saturation intensity Φ_s are generally smaller than the theoretical value of $h\nu/\sigma$ and increase with the output. This fact indicates that σ is not fixed in a glass; this results from heterogeneity of the glass structure. As shown in Fig. 5.18b, a difference in the ligand field surrounding the Nd^{3+} ion results in differences in the distribution of λ, broadening of $\Delta\lambda$, and oscillator strength. The effective wavelength width $\Delta\lambda_{\text{:eff}}$ of the fluorescence indicates the heterogeneity of the glass structure. For energy extraction to be performed efficiently, a low heterogeneity is necessary. Phosphate laser glasses have a lower $\Delta\lambda_{\text{eff}}$ than silicate glasses, and may be characterized as more suitable for energy extraction.

5.2.1.4. Parasitic Oscillation and Clad Glasses

Large-aperture rods or disks are used in high-output laser systems used for nuclear fusion. To maintain the stored energy at a high value, it is necessary to prevent energy loss arising from spontaneous emission and parasitic oscillation arising from reflection at the edges of the glass. For this purpose, it is desirable that σ be small; furthermore, edge cladding is used to prevent parasitic oscillation. In edge cladding, an attempt is made to prevent reflection by cladding the periphery of the disk with a glass that has more or less the same index of refraction as the laser glass and absorbs laser light at 1.06 μm. It is necessary to reduce the reflectance of laser light to approximately 0.1%. For this purpose, glasses doped with Cu^{2+} are used.

5.2.1.5. Nonlinear Coefficient and Self-Focusing

As the intensity of laser light increases, the field strength dependence of the index of refraction can no longer be ignored, and it becomes necessary to consider a nonlinear coefficient n_2 as expressed by the following equation:

$$n = n_0 + n_2|E|^2 \quad . \tag{5.26}$$

When n_2 is large, two things occur. First, as the pulse width decreases and enters the picosecond region, E increases; as a result, the index of refraction of the laser light path area increases, and self-focusing of the laser light occurs as a result of a

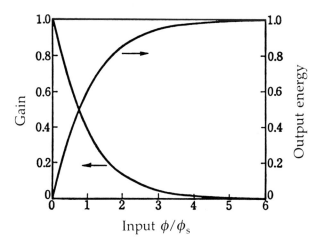

Figure 5.17. Input energy vs output energy, and saturation of gain (Ref. 15).

lens effect, leading to damage. For this reason, the nonlinear coefficient of refractive index of a laser glass, which is used where the pulse width is small, must have a low value. This is the reason that the low-nonlinear-coefficient phosphate-glass LHG 8 (n_2 = 1.13 × 10^{-13} esu) and fluoro-phosphate glass LHG 10 (n_2 = 0.61 × 10^{-13} esu) were developed to replace the silicate laser glass LSG 91H (n_2 = 1.58 × 10^{-13} esu). Second, if there is an intensity distribution in the beam, it is amplified by the nonlinear coefficient so that breakup of the beam occurs, as seen in Fig. 5.19, where the B coefficient, which is defined by the equation shown below, must be 5 or less in a high-power laser device.

$$B = \frac{2\pi}{\lambda} \int_0^l \frac{n_2}{n} I\, dl \quad . \tag{5.27}$$

Here, I is the intensity of the incident light, and l is the length of the medium.

5.2.1.6. Sensitizers

Because the Nd^{3+} absorption band is narrow and the absorption coefficient is small, only a portion of the pumping light from the Xe flash lamp is absorbed. Accordingly, if a sensitizer having a broadband absorption is added and energy transfer is effected from the sensitizer to the Nd^{3+}, the pumping efficiency can be increased. Both Ce^{3+}

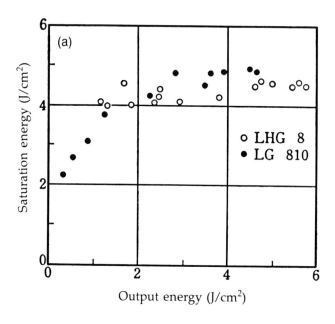

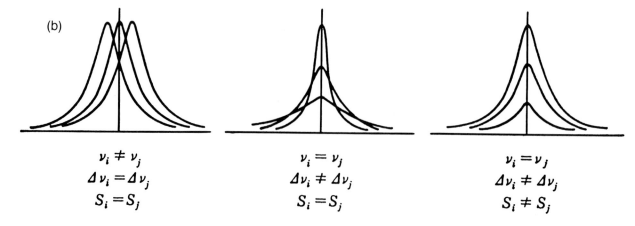

Figure 5.18. (a) Relationship between saturation energy and output energy for LHG 8 and LG 810 (Ref. 15), and (b) graph indicating differences in center frequency, oscillation width and intensity, shown by ions at different sites *i* and *j* (Ref. 17).

and Cr^{3+} have been investigated as sensitizers for Nd^{3+}.

The ion Ce^{3+} shows a broadband absorption in the ultraviolet region, whereas Nd^{3+} shows no absorption; furthermore, the emission band of Ce^{3+} in the vicinity of 350 nm overlaps the absorption band of Nd^{3+}. Accordingly, energy transfer by radiation would be expected to occur. Furthermore, since the fluorescence lifetime of Ce^{3+} decreases with an increase in the concentration of Nd^{3+}, it appears that energy transfer also occurs by nonradiative transitions. Nevertheless, Ce^{3+} has not been observed to have any appreciable sensitizing effect. This may be due to some production of Ce^{4+}, which absorbs the Nd^{3+} pumping light.

The ion Cr^{3+} shows a strong absorption in the visible region, and the emission spectrum of Cr^{3+} overlaps the absorption spectrum of Nd^{3+}. From the decrease seen in the fluorescence lifetime of Cr^{3+} as Nd^{3+} is increased, it appears that nonradiative energy transfer occurs. However, the gain of a glass doped with Cr^{3+} shows no great increase. This may result from the fact that the strength of the Cr^{3+} ligand field in glass differs from that crystals such as GSGG, etc.

5.2.1.7. Solarization

The pumping light may cause solarization, creating color centers so that the absorption efficiency of the Nd^{3+} is lowered. To prevent solarization, the glass may be doped with ions (Ce^{3+}, Nb^{5+}, Mo^{6+}) that absorb the ultraviolet rays that cause solarization; alternatively, the removal of As^{5+} and Sb^{5+}, which stabilize the color centers, is effective in preventing coloring.

5.2.1.8. Concentration Quenching and Effect of Water on the Mean Fluorescence Lifetime

The pumping light absorption efficiency can be improved and ΔN increased by increasing the amount of Nd^{3+} doping. However, nonradiative transitions are increased by concentration quenching so that the emission efficiency drops. The fact that this effect is smaller in phosphate glasses than in silicate glasses is recognized from the dependence of the fluorescence lifetime on the amount of Nd^{3+} doping.

As the Nd^{3+} concentration increases so that excited Nd^{3+} ions are positioned close to unexcited Nd^{3+} ions, direct cross relaxation occurs as a result of dipole-dipole interaction. Furthermore, an energy transfer occurs from the activated ions

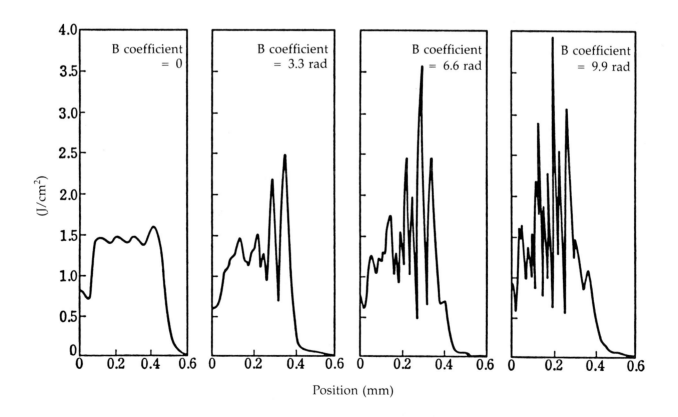

Figure 5.19.   Relationship between beam breakup (intensity fluctuations) and B coefficient.

through the $^4F_{3/2}$ level so that energy is transferred to the acceptors. In this case, the acceptors are either Nd^{3+} ions or (in the case of phosphate glasses) OH^- ions. It is known that in the case of phosphate glasses, the fluorescence lifetime is strongly affected by water. The reason for this is that cross relaxation occurs either directly or after an energy transfer between Nd^{3+} and OH^-. The probability of cross relaxation is given by the following equation:

$$W_{DA} = \frac{C_{DA}}{R^6} \quad . \tag{5.28}$$

Here, W_{DA} is the probability of nonradiative transition loss due to cross relaxation, R is the interionic distance, and C_{DA} is a parameter related to the quenching mechanism. The fact that concentration quenching does not readily occur in the case of phosphate glasses is apparently attributable to the large Nd–Nd distance resulting from the chain-form structure and to the difficulty of Nd–Nd interaction.[18]

Highly doped laser glasses have been developed using the fact that concentration quenching does not readily occur in phosphate glasses. In the case of LHG 8, a glass with 8% doping can put out approximately twice the output of a glass with 3% doping. In this case, the effect of the water content on the mean lifetime of fluorescence is increased, accordingly, the water content must be controlled so that the absorption coefficient K_{OH} of water at $3 \mu m$ is 1 to 3 cm^{-1}.

5.2.1.9. Athermal Glasses for High-Repetition-Rate Lasers

A drawback of using glasses as laser media is that glasses have a low thermal conductivity. Accordingly, when laser oscillation is repeated, the pumping light causes the internal temperature of the rod to rise; as a result, the length of the light path changes and birefringence occurs, and in some cases oscillation may stop.

As mentioned earlier, the temperature coefficient of the light path length dS/dT is given by the following equation:

$$\frac{dS}{dT} = (n - 1)\alpha + \frac{dn}{dT} \quad . \tag{5.28}$$

By selecting a phosphate glass composition in which the temperature coefficient of polarizability is small, it is possible to make dn/dT negative so that dS/dT is approximately zero. In the case of

LHG 8, dS/dT is $+0.6 \times 10^{-6}$, and high repetition rates up to approximately 30 pps are possible; if the glass is highly doped (8%), the output is almost unchanged up to 10 pps. Furthermore, the beam quality shows superior characteristics in both the near-field pattern and far-field pattern. Thus, although the thermal conductivity of the glass cannot be increased; this procedure has made it possible to achieve high-repetition-rate oscillation by decreasing the temperature coefficient of the light-path length.

5.2.2. Laser Glasses for High-Output Applications

High-power laser glasses are basically characterized by σ, τ, and n_2. Table 5.4 shows various characteristics of the silicate laser glass LSG 91H; phosphate laser glasses LHG 5, LHG 7 and LHG 8, and fluorophosphate laser glass LHG 10.

5.2.3. Physical Properties of Laser Glasses (Relationship between Coefficient of Stimulated Emission and Composition)

We might say that the goal of research on the physical properties of laser glasses is to clarify the relationship between the coefficient of stimulated emission and the structure of the glass, or to find out what determines the radiative transition probability and nonradiative transition probability. Some especially interesting questions follow: Why do phosphate laser glasses show a higher coefficient of stimulated emission than silicate laser glasses? In addition, the effects of alkali ions on radiative transitions and nonradiative transitions appear in exactly the reverse order in the two glass types when ionic radius is taken as a parameter. What is the reason for this?

5.2.3.1. Radiative Characteristics

Radiative transitions in Nd^{3+} are based on f–f electron transitions, and the electric-dipole transitions are forbidden. If the ligand field is asymmetrical, however, the parity becomes odd so that transitions are allowed. Accordingly, the covalent bonding nature of the Nd–O bonds and the asymmetry of the ligand field contribute to the transition probability.

The intensity parameter Ω_t given by Judd-Ofelt's theory can be expressed (as shown in the

Table 5.4. Characteristics of Hoya laser glasses.

	LSG 91H	LHG 5	LHG 7	LHG 8	LHG 10
LASER CHARACTERISTICS					
Nd_2O_3 (wt%)	3.1	3.3	3.4	3.0	2.4
Nd^{3+} ion concentration (10^{20} ions/cm^3)	3.0	3.2	3.1	3.1	3.1
Cross-sectional area of stimulated emission, $\sigma_P(10^{-20}$ cm^2)	2.7	4.1	3.8	4.2	2.7
Fluorescence lifetime (μs)	300	290	305	315	384
Gain coefficient (cm^{-1} J/cm^3)	0.144	0.217	0.202	0.223	0.143
Half-width, $\Delta\lambda$ (290 K)(Å)	274	220	222	218	265
Center wavelength, λp (nm)	1062	1054	1054	1054	1051
Attenuation coefficient (1054 nm)(m^{-1})	0.1 (1062 nm)	0.123	0.131	0.1	0.15
Slope efficiency (%) ($10\phi \times 160\ l$ mm, R = 80%)	1.15	1.83	1.82	1.83	
Oscillation threshold (J) ($10\phi \times 160\ l$ mm, R = 80%)	52	40	35	32	
Relative performance index, σ/n_2	1.00	1.87	2.12	2.17	
LASER DAMAGE THRESHOLD					
Internal damage (30 ns pulse) (J/cm^2)	400	400	400		
Surface damage (30 ns pulse) (J/cm^2)	25	50	80		
OPTICAL CHARACTERISTICS					
Nonlinear index of refraction, n_2 (10^{-13} esu) (calculated value)	1.58	1.28	1.05	1.13	0.61
Index of refraction					
n (1054 nm)	1.54980 (1062 nm)	1.53078	1.50415	1.52005	1.46078 (1051 nm)
n (632.8 nm)	1.55901	1.53909	1.51159	1.52793	1.46602
n_d	1.56115	1.54096	1.51316	1.52962	1.46715
n_L	1.56804	1.54686	1.51843	1.53515	1.47084
n_C	1.55812	1.53834	1.51083	1.52718	1.46551
Abbe number, ν_d	56.56	63.49	67.56	66.49	87.68
Brewster's angle	57°10′	56°50′	56°22′	56°40′	55°36′
Temperature coefficient of index of refraction (10^{-6}/°C) (20 to 40°C)	+1.6	0.0	-2.9	-5.3	-7.7
Coefficient of linear expansion (10^{-6}/°C) (20 to 40°C)	9.0	8.6	10.2	11.2	14.5
Temperature coefficient of light-path length (10^{-6}/°C) (20 to 40°C)	+6.6	+4.6	+2.3	+0.6	-1.0
THERMAL CHARACTERISTICS					
Yield point, Ts (°C)	505	486	543	520	470
Transition point, Tg (°C)	465	455	510	485	445
Coefficient of linear expansion (100 to 300°C) (10^{-7}/°C)	105	98	112	127	153
Thermal conductivity (25°C) (kcal/m·h·°C)	0.89	0.66	0.62	0.50	0.64
Specific heat (cal/g·°C)	0.15 (50°C) 0.17 (122°C) 0.19 (246°C)	0.17 (20°C) 0.18 (60°C) 0.20 (240°C)	0.17 (25°C) 0.18 (50°C) 0.22 (250°C)	0.18 (25°C) 0.19 (50°C) 0.24 (250°C)	
CHEMICAL CHARACTERISTICS					
Water resistance (wt. loss %) (H_2O, 100°C, 1 hr)	0.036	0.08	0.13	0.13	0.034
Acid resistance (wt. loss %) (HNO_3 pH 2.2, 100°C, 1 hr)	0.039	0.16	0.38	0.47	0.337
OTHER CHARACTERISTICS					
Density (g/cm^3)	2.81	2.68	2.60	2.83	3.64
Young's modulus (kgf/mm^2)	8890	6910	5640	5110	7250
Shear modulus (kgf/mm^2)	3590	2790	2780	2030	2790
Poisson's ratio	0.237	0.237	0.238	0.258	0.300
Knoop hardness (100 g) (kgf/mm^2)	590	497	367	321	361
Photoelastic constant (nm/cm^{-1}·kgf^{-1}/cm^2)	2.16	2.26	2.17	1.93	

following equation) as a function of A_{tp}, which indicates the asymmetry of the Nd^{3+} ligand field, and Ξ, which indicates the covalent bonding nature[19]:

$$\Omega_t \propto |A_{tp}|^2 \Xi^2, t = 2,4,6 \quad . \tag{5.29}$$

The term Ξ can be estimated from the nephelauxetic parameter β (Ref. 20), which indicates the magnitude of electron donation from the coordinate oxygen ions to Nd^{3+}. Furthermore, β can be determined in relation to optical basicity from the following equation:

$$\beta = \frac{v_t - v}{v_t} \quad . \tag{5.30}$$

Here, v_t is the absorption wave number of the probe ions in the free-ion state, and v is the absorption wave number of the probe ions in the glass. The Pb^{2+} ions were used as probe ions. As shown in Fig. 5.20b, the value of β determined from the shift in the absorption wavelength of Pb^{2+} in the various glasses is larger in silicate glasses than in phosphate glasses; it thus appears that the covalent bonding nature (Ξ) of Pb^{2+}–O bonds is less in phosphate glasses than in silicate glasses. Accordingly, the large A_{rad} values seen in phosphate glasses do not depend on the covalent bonding nature of the Nd^{3+}–O bonds, but rather on the asymmetry of nonpolar transitions of the Nd^{3+} ligand field. When asymmetry was expressed as the emission intensity ratio of electronic dipole transitions to magnetic dipole transitions of Eu^{3+} (Fig. 5.20c), the above fact was confirmed by the comparison between this ratio and the radiative transition probability.[21] The higher asymmetry of phosphate glasses results from the fact that phosphate glasses have a two-dimensional, chain-form structure, whereas silicate glasses have a three-dimensional network structure (see Fig. 5.21).

The change in A_{rad} that occurs when the type of cation is changed can be explained as follows: In silicate glasses, the asymmetry of the Nd^{3+} ligand field increases with an increase in the cationic field strength. In phosphate glasses, on the other hand because the cations play a role in connecting chains to each other, the asymmetry of the ligand field surrounding Nd^{3+} increases as the cationic field becomes weaker.

5.2.3.2. Nonradiative Characteristics

Nonradiative transitions occur through (1) concentration quenching, (2) energy transfer, and (3) multiphonon relaxation. In cases where the Nd^{3+} ion concentration is low and no OH ions are present, nonradiative transitions occur by a process of multiphonon relaxation (Fig. 5.22). The probability is given by the Miyakawa-Dexter equations:

$$W_p = W_0 \exp(-\alpha \Delta E) \quad . \tag{5.31}$$

$$\alpha = (\hbar \omega)^{-1}\left(\ln \frac{p}{g} - 1\right) \quad . \tag{5.32}$$

Here, p is the number of phonons, g is the electron-phonon coupling constant, ΔE is the energy gap, and $\hbar \omega$ is the phonon energy.

Weber et al.,[22] showed that W_p increases in the order of tellurite, germanate, silicate, phosphate, and borate glasses, which agrees with the order of the maximum vibrational energies of the Raman scattering spectra of the respective glasses. Weber et al.,[22] also found that W_p is determined by the maximum phonon energy $\hbar \omega$ of the glass network involved. However, W_p also varies according to the modifier ions used. Toratani et al.,[21] measured g and $\hbar \omega$ and determined W_p from measurements of phonon sidebands in excitation spectra of Eu^{3+}. As a result, it was found that the effect of modifier components follows the variation of g (indicating that modifier components affect the electron-phonon coupling constant), and that the effect of network-forming components appears in the phonon energy (see Fig. 5.22).

5.2.4. Laser-Related Glasses

5.2.4.1. Faraday Rotation Glasses

In laser fusion systems, amplified laser light is directed against a target. However, there is a danger that some of the light will be reflected by the target and will travel back and destroy the laser disk or other optical components. To prevent this, an optical isolator is needed that allows the passage of light only in the forward direction and prevents the passage of light in the backward direction.

Faraday rotation glasses are used as optical isolators. Laser light that has been linearly polarized by a polarizer on the input side has its plane of polarization rotated 45 deg by a Faraday rotation glass placed in a magnetic field. When the same laser light returns after being reflected by the target, it again passes through the Faraday rotation glass so that its plane of polarization is rotated a further 45 degrees. As a result, the plane of

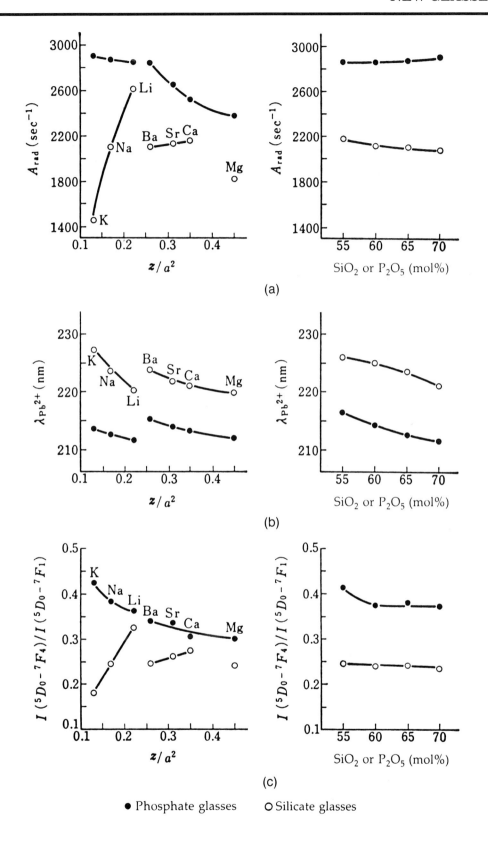

● Phosphate glasses   ○ Silicate glasses

Figure 5.20. (a) Radiative transition probability A_{rad} of silicate glasses and phosphate glasses, (b) absorption wavelength of Pb^{2+} probe ions, and (c) asymmetry.

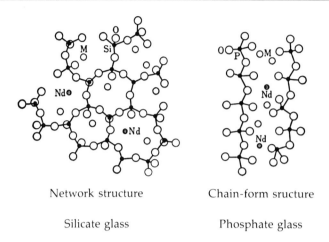

Figure 5.21.   Structural models of silicate and phosphate glasses.

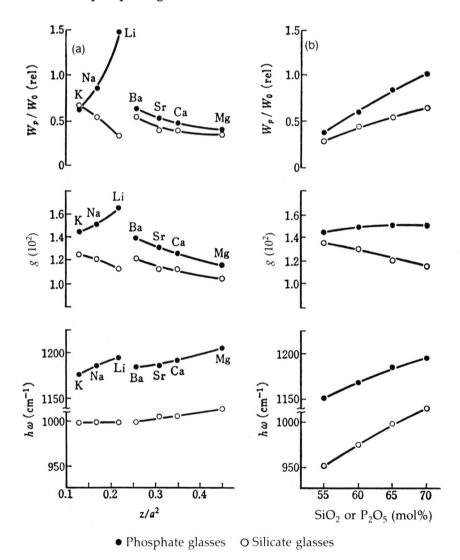

● Phosphate glasses   ○ Silicate glasses

Figure 5.22.   Variation of multiphonon relaxation rate, phonon energy, and electron–phonon coupling constant with (a) ionic field strength, and (b) modifier oxide concentration in silicate and phosphate glasses.

polarization of the light becomes perpendicular to the plane of polarization of the polarizer on the input side, and the light cannot pass through. The relationship between the rotation angle of linear polarization and the intensity of the magnetic field is given by the following equation:

$$\theta = VHl \quad . \tag{5.33}$$

Here, θ is the rotation angle, l is the length of the substance, and H is the intensity of the magnetic field. The variable V is a proportional constant called the Verdet constant (min/Oe·cm), which is specific to the substance involved. In the case of diamagnetism, this constant is positive; in the case of paramagnetism, it is negative. The following requirements must be met in Faraday rotation glasses used in laser fusion devices: the Verdet constant must be large, but the absorption at 1.06 μm must be low; in addition, the nonlinear optical constant must be small and the glass must be a glass that can be obtained in a large, homogeneous form.

The cause of the Faraday effect is as follows: the intrinsic magnetic moments of the constituent atoms of the magnetic medium are arranged in the direction of the magnetic field by the action of the magnetic field so that the dispersion curves for left and right circularly polarized light are different. The rates of travel of left and right circularly polarized light are different in a substance placed in a magnetic field so that the respective light beams are out of phase when they have passed through the substance. Accordingly, the plane of polarization of the reconstituted linearly polarized light is rotated. The Verdet constant of a dia magnetic glass can be expressed by Bequerel's equation[23] in which only the normal Zeeman effect is considered

$$V = \left(\frac{vg}{2mc^2}\right)\left(\frac{dn}{dv}\right) \quad . \tag{5.34}$$

Here, v is the frequency of the light, and g is the Landé factor. The second term in this equation represents dispersion, and indicates that the Verdet constant increases with an increase in dispersion. This indicates that high-dispersion glasses containing large numbers of Tl^+, Pb^{2+}, Bi^{3+}, or Te^{4+} ions will show a high Verdet constant. On the other hand, the Verdet constant of a paramagnetic glass containing rare-earth ions that have unpaired electrons can be expressed by the following equation derived by van Vleck et al.,[24]

$$V = \frac{(4\pi)^2 \mu_B v^2}{3chkT}\frac{Np}{g}\sum\frac{c_n}{v^2 - v_n^2} \quad . \tag{5.35}$$

Here, μ_B is the Bohr magneton, N is the number of paramagnetic ions per unit volume, p is a quantity related to the effective magneton ($p = g\sqrt{J + 1}$), J is the total angular momentum quantum number, v_n is the inherent absorption frequency, and c_n is the transition probability. The Σ term in the equation represents dispersion; accordingly, a glass with a high Verdet constant can be obtained by including large numbers of ions with a large p value and a high dispersion in the glass. Among rare-earth elements, glasses show a high Verdet constant when they contain Ce^{3+}, which shows an inherent absorption in the near ultraviolet region, or when they contain Tb^{3+}, Dy^{3+} or Eu^{2+}, which have a large p value. Figure 5.23 shows the Verdet constants of phosphate glasses containing rare earths.

Paramagnetic glasses show a higher Verdet constant than diamagnetic glasses. Table 5.5 shows various characteristics of diamagnetic and paramagnetic Faraday-rotation glasses. Here, V/α and $V(n/n_2)$ are used as figures of merit.

5.2.4.2. Antireflection Coating ARG 2

Large numbers of optical elements are used in a high-output laser; consequently, spatial filters and focusing lenses, etc., are coated with antireflection coatings. However, in the case of antireflection coatings formed by conventional vacuum evaporation, the laser power is restricted because the laser breakdown threshold is low, i.e., approximately 5 J/cm². Accordingly, an antireflection film that is able to resist laser breakdown needs to be developed. The antireflection film ARG 2 was developed in response to this demand.

The method used to produce the antireflection film consists of forming a porous leached layer on the surface of a borosilicate glass by the acid treatment (under appropriate conditions) of the surface of the glass after the glass has been subjected to a phase-separation heat treatment. This method was first used by Minot[25] on pyrex glass. When heat-treated, a borosilicate glass separated into (1) an SiO_2-rich phase that does not readily dissolve in acid, and (2) a phase rich in alkali and boric acid that readily dissolves in acid. When this glass is immersed in a solution formed by adding a small amount of ammonium fluoride to nitric acid, the easily-dissolved phase is selectively leached so that an SiO_2-rich layer with a porous structure remains on the surface of the glass. Asahara et al.,[26] applied this method to optically homogeneous ARG 2 glass and obtained the spectroscopic reflectance curve shown in Fig. 5.24. Asahara et al.,[26] also performed an optical structural analysis of a thin film formed in this way, and found that this thin film has an index-of-refraction gradient: the index of refraction of the

film in contact with air is close to that of air, whereas the index of refraction of the film in contact with the glass is close to that of the bulk glass. A structural model of such a film is shown in Fig. 5.25. Asahara et al.,[26] determined the index-of-refraction $n_f(a)$ of the film in contact with the air, the index of refraction $n_f(g)$ of the film in contact with the glass, and the thickness d_f of the film from the maximum (R_{max}) and minimum (R_{min}) of

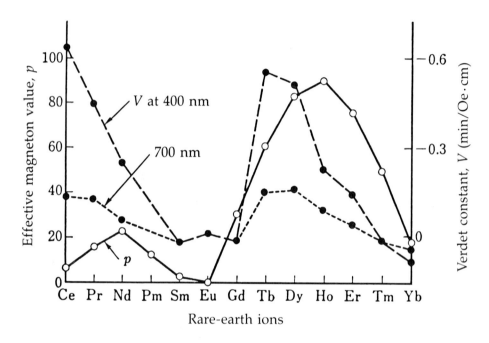

Figure 5.23.   Effective magneton values and Verdet constants of phosphate glasses containing rare earths.

Table 5.5   Various characteristics of Faraday rotation glasses.

	Diamagnetic glass	Paramagnetic glasses	
	SF 6	FR 4 (Hoya)	FR 5 (Hoya)
Verdet constant (min/Oe·cm)			
0.633 μm	0.093	−0.104	−0.268
1.06 μm	0.028	−0.035	−0.082
Absorption coefficient			
α (1.06 μm)	0.0065	0.0054	0.0085
n_d	1.8052	1.5718	1.6862
v_d	25.4	56.8	53.2
n (1.064)	—	1.556	1.678
n_2 (10^{-13} esu)	9.9	1.8	2.5
Performance indices			
V/α	4.3	6.5	9.6
$V(n/n_2)$	0.005	0.030	0.055

the reflection spectrum shown in Fig. 5.24, using the following equations:

$$\left[n_f(a) + n_f(g)\right] d_f$$

$$= 2m \cdot \frac{\lambda}{2} : R_{\min} = \frac{\left[n_f(a)n_g - n_f(g)\right]^2}{\left[n_f(a)n_g + n_f(g)\right]^2} \quad . \qquad (5.36)$$

$$\left[n_f(a) + n_f(g)\right] d_f$$

$$= (2m + 1) \cdot \frac{\lambda}{2} : R_{\max} = \frac{\left[n_f(a)n_f(g) - n_g\right]^2}{\left[n_f(a)n_f(g) + n_g\right]^2} \quad . \qquad (5.37)$$

Here, n_g is the index of refraction of the glass. When $R_{\max} = 0.33\%$, $R_{\min} = 0.12\%$, and $n_g = 1.489$ were substituted into the equations, values of $n_f(a) = 1.10$ and $n_f(g) = 1.452$ were obtained. Furthermore, when $\lambda_{\max} = 1.07$ μm and $\lambda_{\min} = 0.955$ μm were substituted into the equations, a value of $d_f = 1.68$ μm was obtained. Furthermore, assuming that the index of refraction n_{us} of the SiO$_2$–rich phase is 1.46, the volume fraction of the residual SiO$_2$ phase is calculated as $V_a = 24\%$ on the air side, and as $V_g = 98\%$ on the glass side, from the following equation:

$$V = \frac{\left(n_f^2 - 1\right)\left(n_{us}^2 + 2\right)}{\left(n_f^2 + 2\right)\left(n_{us}^2 - 1\right)} \quad . \qquad (5.38)$$

This indicated that the gradient refractive index layer has a high porosity on the air side.

Nitric acid and acidic ammonium fluoride were used in the acid treatment solution. Nitric acid alone can form an index-of-refraction gradient by selective leaching; however because this acid causes a weak corrosion, it cannot form a film that has a sufficiently high porosity on the air side. On the other hand, acidic ammonium fluoride can form a film with a high porosity; however because this substance causes a strong and uniform corrosion to occur, an index-of-refraction gradient is not formed. Only by using a mixed solution of both substances an index-of-refraction gradient film with a high porosity can be formed. In this case, the acidic ammonium fluoride plays the role of increasing the porosity of the air-side layer, and thus increasing the index-of-refraction gradient formed by the nitric acid. The conditions of this process are shown in Fig. 5.26. The reflectance of a thin film formed in this way showed almost no wavelength dependence and also showed a value of less than 0.5% in the wavelength range from 0.4 to 1.4 μm. Furthermore, the

laser damage threshold showed a mean value of 12 J/cm^2, i.e., a value approximately 2.5 times higher than the value shown by a conventional vacuum-evaporated film. However, considerable surface scattering was observed.

A second method for obtaining an antireflection film with a high laser damage threshold employs a sol-gel reaction.[27] This method uses the fact that a structure with a high porosity is formed in the sol-gel reaction process, but the difficulty with this method is to coat a large disk with a film of uniform thickness.

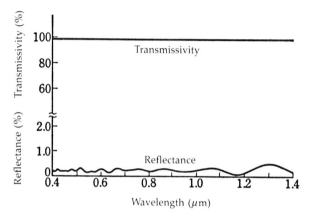

Figure 5.24.   Reflection and transmission spectra of ARG 2 (antireflection glass).

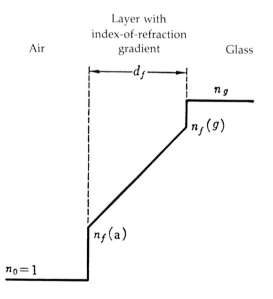

Figure 5.25.   Model showing index-of-refraction gradient of antireflection layer.

A third method for forming a thin film resistant to laser damage is a method called the "Schroeder method." This method attempts to lower the reflectance by deliberately forming a glass staining layer on the surface of the glass. Easily dissolved components in the glass are leached while the pH is maintained at a fixed level by means of a weakly alkaline solution (sodium arsenite) so that a porous surface layer consisting of a silica framework is formed. Because the glass does not undergo phase separation in this case, the pore diameter is small (70 to 100 Å), and the film is thin. The reflectance shows a wavelength dependence, even though the film has an index-of-refraction gradient. This effect is caused by a large difference between $n_f(g)$ and n_g, and the film to some extent displays the properties of a film with a single refractive index. The breakdown threshold shows a fairly high value,

i.e., 9 J/cm^2 (400 ps).[28] Table 5.6 shows the characteristics of antireflection films manufactured by each of the methods described above. All of the films show a weaker adhesion than vacuum-evaporated films.

5.2.4.3. Frequency-Conversion Filters

It has been discovered that as the wavelength of laser light grows shorter, absorption by a laser-fusion target increases abruptly, backscattering of hot electrons decreases, and ablatively-driven contraction of the target more readily occurs.[29] As a result, the wavelengths of lasers presently in use are in the process of being shifted from the current 1.06 μm to the second and third harmonics (0.53 and 0.355 μm, respectively). What becomes necessary in this case is a filter that allows only the second or third harmonic wave to pass

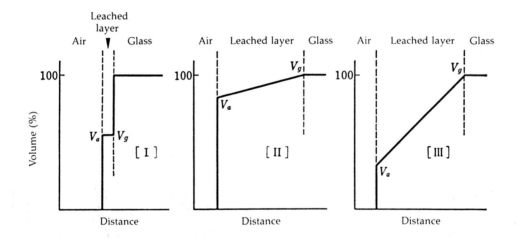

Figure 5.26. Volume % in various layers when acid treatment was performed with different solutions: (I) acidic ammonium fluoride solution, (II) nitric acid solution, (III) mixed solution of nitric acid and acidic ammonium fluoride.

Table 5.6. Various antireflection films and their properties.

	Reflectance (%)		Breakdown threshold (J/cm^2)			
	0.53 μm	1.06 μm	0.35 μm	0.53 μm	1.06 μm	Thickness (μm)
Elution of phase-separated glass	0.25	0.25	10.1 (700 ps)	12.5 (700 ps)	19.5 (1 ns)	1 to 2
Sol-gel method		0.15 to 0.4			19 to 23 (1 ns)	0.14
Schroeder method		0.01	9 (400 ps)		8 to 15 (1 ns)	0.21

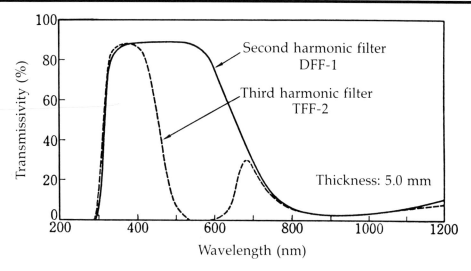

Figure 5.27. Transmissivity curves of second harmonic and third harmonic filters.

through, and that absorbs light at other frequencies. The 2ω- and 3ω-frequency-conversion filters, such as those shown in Fig. 5.27 and Table 5.7, have been developed by introducing CuO or CoO, NiO, etc., which absorb light at 1.05 and 0.53 μm, into a suitable basic glass. In this case, it is necessary to avoid solarization caused by multiphoton absorption. Furthermore, a filter has also been developed that absorbs second and third harmonic waves and allows only 1.06-μm light to pass through.

Table 5.7. Characteristics of 2ω- and 3ω-frequency-conversion filters.

Wavelength (nm)	Transmissivity (%)	
	DFF (2ω)	TFF (3ω)
1060	4.5	4.5
530	88.5	4.5
353		87.2

5.3. Acousto-Optical Glasses

Acousto-optical control devices that use ultrasonic waves to deflect or modulate light take advantage of the fact that periodic variations in index of refraction generated in a transparent medium by ultrasonic waves act as a diffraction grating for light so that the light is diffracted. Special features of such systems are that they are smaller than electrical light modulation devices, achieve a high extinction ratio at a low driving power, and are inexpensive. It has been found that tellurite glasses are useful as ultrasonic wave media in such acousto-optical devices.[30]

When light is Bragg-reflected by a diffraction grating created by ultrasonic waves, the deflection angle θ and intensity of the diffracted light I are given by the following equations (as shown in Fig. 5.28):

$$\sin \theta = \frac{K}{k} \quad . \tag{5.39}$$

$$I = I_0 \sin^2 \left(\frac{A}{\lambda^2} M_e P \right)^{1/2} \quad . \tag{5.40}$$

Here, K and k are the wave numbers of the ultrasonic waves and the laser light, respectively; P is the ultrasonic wave power; λ is the wavelength of the laser light; A is a factor determined by the shape of the ultrasonic beam; and M_e is a physical constant that is peculiar to the substance involved and is called the "figure of merit." The direction of the diffracted light (deflection angle) and the intensity of the diffracted light are determined by the wave number and power of the ultrasonic waves.

Acousto-optical materials must satisfy the following conditions: (1) the figure of merit must be high; (2) there must be little ultrasonic absorption; and (3) the light transmission wavelength region must be broad.

As shown by Eq. (5.40), the intensity of the diffracted light is determined by the power of the ultrasonic waves, the figure of merit of the medium, the shape factor of the ultrasonic beam, and the wavelength of the light. The figure of merit is determined by the characteristics of the medium and is given by the following equation:

$$M_e = \frac{n^6 p^2}{\rho v^3} \; .\tag{5.41}$$

Here, n is the index of refraction, p is the photoelastic constant, ρ is the density, and v is the velocity of sound. A high figure of merit signifies that only a small electrical input is required for modulation and deflection. As shown by Table 5.8, the tellurite glasses AOT 5 and AOT 44B developed at Hoya Glass have a high figure of merit.

Tellurite glasses show a high figure of merit because these glasses have a high index of refraction. In addition, the fact that the velocity of sound is low in such glasses also makes some contribution to this high figure of merit. Moreover,

the tellurite glass AOT 5 has a relatively low ultrasonic absorption coefficient. The ultrasonic absorption mechanism of this glass has been studied by Izumitani et al.[31] Ultrasonic absorption by tellurite glasses shows the following characteristics:

1. The frequency dependence of absorption is as shown in Fig. 5.29; absorption is proportional to approximately the square of the frequency in the range from 10 to 200 MHz.

2. Absorption is independent of temperature in the range from −200 to 150°C (Fig. 5.30).

3. The absorption coefficient is proportional to the relaxation time determined from the thermal conductivity.

Accordingly, ultrasonic absorption by tellurite glasses in these regions depends on interaction

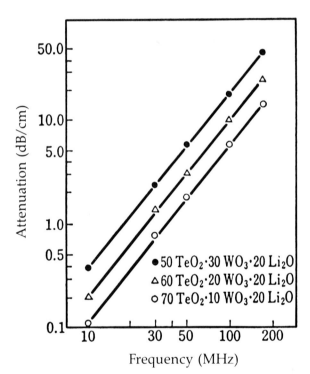

Figure 5.29. Frequency dependence of ultrasonic absorption by tellurite glasses at room temperature.

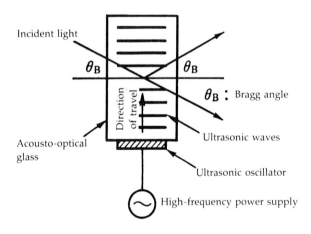

Figure 5.28. Principle of the acousto-optical effect.

Table 5.8. Properties of tellurite glasses AOT 44B and AOT 5.

	AOT 44	AOT 5	SF 6
Index of refraction, n (633 nm)	1.971	2.090	1.7989
Velocity of sound (cm/s)	3.33×10^5	3.40×10^5	3.53×10^5
Figure of merit (6328 Å) (s/g)	20×10^{-18}	24×10^{-18}	4.15×10^{-18}
Ultrasonic absorption (100 MHz) ($\alpha\beta$/cm)	2.0	3.0	—
Light transmission wavelength region (μm)	0.43 to 2.7	0.47 to 2.7	—

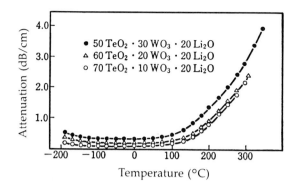

Figure 5.30. Temperature dependence of ultrasonic absorption by tellurite glasses at 10 MHz.

between the lattice vibration phonons and the ultrasonic wave phonons at $\omega\tau \ll 1$, and can be expressed by Woodruff-Ehrenrich's equation[32]:

$$\alpha = \frac{c_V T \gamma^2}{3\rho V^3} \omega^2 \tau_l \quad . \tag{5.42}$$

Here, c_V is the specific heat, γ is the mean Grüneisen constant, v is the velocity of sound, ρ is the specific gravity, ω is the angular frequency, and τ_l is the relaxation time of the lattice phonons. The above equation indicates that ultrasonic absorption is independent of temperature, that α is proportional to τ_l because the ultrasonic loss is proportional to ω^2, and that τ_l is proportional to the reciprocal of the temperature.

Tellurite glasses show a high figure of merit and a low ultrasonic loss. The reason for this is that the network-forming oxide TeO_2 (which imparts a high index of refraction) constitutes the major part of the composition, whereas the amount of modifying oxides (Li_2O, WO_3) is small so that the relaxation time τ_l of the lattice phonons is relatively short. Low ultrasonic absorption means that higher-frequency ultrasonic waves can be used; this means that the ultrasonic bandwidth can be broadened, and that the resolving power and the deflection angle can be increased.

The laser light used in acousto-optical devices ranges from the short-wavelength light of He–Cd lasers, etc., to the infrared light of semiconductor lasers and Nd lasers, etc. The use of short-wavelength laser light such as He–Cd laser light (441.6 nm) makes it possible not only to increase the deflection efficiency, but also to increase the efficiency of peripheral devices such as light-receiving parts, etc. For example, a switch from He–Ne laser light to He–Cd laser light results in a doubling of the deflection efficiency. The glass AOT 44 is completely transparent in the range from 440 to 2700 nm and was developed for use with He–Cd lasers. On the other hand, because tellurite glasses are high-refraction glasses, they show a high Fresnel reflection (12%). However, by coating such glasses with an SiO_2 antireflection coating, it is possible to increase the transmittance to 96%.

5.4. Delay-Line Glasses

Delay-line glasses are glasses where the velocity of an elastic wave in a glass is slower than the velocity of an electromagnetic wave; the glass, therefore, can be used to delay or accumulate electrical signals (for a certain period). As shown in Fig. 5.31, the operating principle is as follows: the electrical signal that needs to be delayed is converted into an elastic wave by a transducer on the input side and is then reconverted into an electrical signal by a transducer on the output side after passing through the glass. Because the velocity of an elastic wave is approximately $1/10^5$ slower than the velocity of an electromagnetic wave, the signal is accumulated and delayed by the time it spends as an elastic wave. Around 1960, the PAL and SECAM systems were employed as color television transmission systems in Europe. In both of these systems, a 1-H (i.e., 64-μs) delay line is required in the image receiver.

Accordingly, a glass delay line suited to this delay time (64 μs) was sought. Recently, with the increase in the recording density of home video tape recorders (VTR), delay-line glasses have begun to be used to eliminate color signal crosstalk. Here, therefore, we will discuss (1) applications and necessary characteristics of delay-line glasses, and (2) physical properties of delay-line glasses.

5.4.1. Applications and Necessary Characteristics of Delay-Line Glasses

5.4.1.1. Application to Color Television

In a color television broadcast, data obtained from the image are divided into three signals, i.e., a signal that indicates brightness (black and white) and two color signals (red and blue) that represent image color data. The brightness signal is broadcast so that it can be received as a black and white signal by black and white television receivers. For color receivers, the green color signal is expressed as a remainder obtained by subtracting the red and blue color signals from the brightness signal. There are three currently used color television broadcasting systems: i.e., the NTSC system, which is used in Japan and the United States; the SECAM system, which is used in the Soviet Union and eastern Europe; and the PAL system, which is used in western Europe. In each of these systems, the method used to transmit and receive these two different color signals is different.

NTSC System. The two different color signals are phase modulated and transmitted independently and simultaneously. The receiver only needs to receive these signals and synthesize them "as is" to reproduce the color of the image.

SECAM System. The two different color signals are alternately transmitted every scanning line period (64 μs) using a frequency modulated subcarrier. Accordingly, a 1-H delay line that accumulates (for 64 μs) the color signal received at a given instant and superimposes it on the color signal received subsequently is required to reproduce the color of the image in the receiver.

PAL System. The two different color signals are phase inverted at each scanning line and transmitted independently. At the receiver, each different color signal is delayed by one scanning line and added to the subsequent signal so that the phase shift is corrected, and image quality is improved. A 1-H delay line is required for this purpose.

5.4.1.2. Video Tape Recorder (VTR) Applications

In home VTR recording systems, guard bands have been installed in the past to prevent

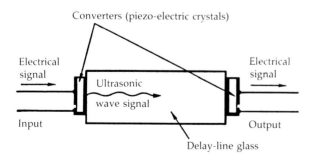

Figure 5.31. Operating principle of ultrasonic delay line.

crosstalk from neighboring tracks. Now, however, a high-density system has been developed that dispenses with these guard bands. The color-signal crosstalk is eliminated by recording a color signal with its phase "as is" on track A, and recording a color signal with its phase inverted 180 deg at each horizontal scanning line on track B (see Fig. 5.32). When track A is reproduced, crosstalk from track B can be cancelled by adding a component that has passed through a 1 H delay line.

5.4.1.3. Miscellaneous

In addition to the uses described above, other VTR-related uses of glass delay lines include the use of a 1-H delay line to compensate for data dropout caused by tape scratches, and the use of a 1/2 H delay line for fast-forward and still functions. Recently, glass delay lines have also been used for color reproduction in video disks.

5.4.2. Necessary Characteristics of Delay-Line Glasses

The following characteristics are required in glasses used as delay media:

1. The transverse-wave sound velocity must be low; i.e., it must be possible to obtain a long delay time with a short glass length so that the delay line can be made compact.

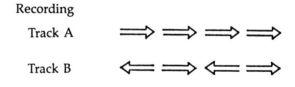

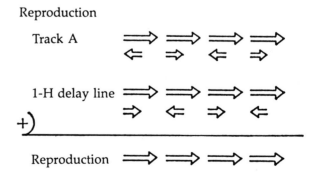

Figure 5.32. Principle of crosstalk prevention using a delay line.

2. The glass must be free from bubbles and veins to prevent scattering of the ultrasonic waves.

3. The transverse-wave attenuation must be small so that the output can be increased.

4. The delay time must not vary with temperature in the temperature range of ordinary use (-20 to $60°C$); i.e., the temperature coefficient of the delay time (TCDT) must be close to zero.

5. The delay time must not vary over time.

Table 5.9 shows the required TCDT characteristics and tolerance ranges for delay-line glasses used in various applications. In the SECAM system, a variation of ± 50 ns in a delay time of 64 μs is permitted because a certain amount of variation in the color signals will remain undetected by the human eye. In the case of PAL, on the other hand, a phase deviation of 8 deg will be seen as color fringes when viewed from a distance of one meter from the receiver. Accordingly, at a frequency of 4.43 MHz, ± 5 ns (which corresponds to ± 8 deg) is considered to be the limiting value of phase deviation. Accordingly, the TCDT must be within ± 5 ns. In VTRs, a standard similar to that used in PAL is applied in the case of crosstalk prevention; for dropout compensation, however, the standard is fairly loose, i.e., a value within ± 100 ns is permissible.

5.4.3. Physical Properties of Delay-Line Glasses

5.4.3.1. Temperature Characteristic of the Delay Time

The delay time (τ) can be expressed as the reciprocal of the sound velocity (v), and is related as follows to the shear modulus (G) and density (ρ) of the glass:

$$\tau = \frac{l}{v} = l \sqrt{\frac{\rho}{G}} \quad . \tag{5.43}$$

Accordingly, the temperature coefficient of the delay time is

$$\frac{1}{\tau}\frac{d\tau}{dT} = -\frac{1}{2}\left(\frac{1}{l}\frac{dl}{dT} + \frac{1}{G}\frac{dG}{dT}\right) \quad ,$$

$$= -\frac{1}{2}(\alpha + \gamma) \quad , \tag{5.44}$$

and can be expressed as the sum of the expansion coefficient α and the temperature coefficient γ of the shear modulus. Here, α ranges from 50 to 100×10^{-6}, whereas γ varies markedly with the composition of the glass, ranging from the 144×10^{-6} of SiO_2 at the high value to -100×10^{-6} at the low value (Ref. 33). Consequently, for the TCDT to be close to zero, it is necessary that γ show a negative value of the same order as α. The value l varies almost linearly with the temperature, and the value of α is more or less fixed with respect to temperature. The shear modulus, on the other hand, can be expressed as an approximate second-order curve[34] that decreases with temperature. Accordingly, in the temperature region where α and γ are of the same order,

$$|\alpha| > |\gamma|, \quad \text{and} \quad \frac{1}{\tau}\frac{d\tau}{dT} < 0$$

in the low-temperature range, and

$$|\alpha| < |\gamma|, \quad \text{and} \quad \frac{1}{\tau}\frac{d\tau}{dT} > 0$$

in the high-temperature range. Accordingly, the temperature characteristic of the delay time can be expressed as a second-order curve with a minimum at the value of

$$\frac{1}{\tau}\frac{d\tau}{dT} = 0 \quad .$$

To minimize the TCDT in the vicinity of room temperature, the composition is selected so that this minimum occurs near room temperature.

Among glass components, only SiO_2 produces a negative TCDT. Therefore, to obtain a glass with a TCDT near zero, the TCDT is adjusted by adding modifying oxides that produce a

Table 5.9. Necessary characteristics of delay-line glasses.

| | Color television | | VTR | | | Video disk, |
	PAL	SECAM	Crosstalk filter	Dropout compensation	Fast forward	VHD
Delay time	1 H	1 H	1 H	1 H	1/2 H	1 H
Frequency (MHz)	4.43	4.43	3.58	3.58	3.58	3.58
TCDT (ns/64 μs)	± 5	± 50	± 5	± 100	± 5	± 5

positive TCDT. Figure 5.33 shows compositions in which the TCDT is zero in the case of a ternary system of SiO_2–PbO–K_2O (Refs. 35, 36). Compared with commercially marketed optical glasses, these compositions are rich in SiO_2. Both PbO and K_2O are used as modifying oxides not just because they are easily-dissolved components, but also because PbO in particular has a two-fold advantage: it decreases the sound velocity and because its relative effect in making the TCDT positive is small, it can be introduced into the glass in large amounts.

5.4.3.2.   Effect of Heat Treatment on the TCDT

As shown in Fig. 5.34, the temperature characteristic curve of the delay time varies with the cooling rate. Compared with slow cooling, rapid cooling causes a shift of the minimum value of $d\tau/\tau$ toward lower temperatures. This can be interpreted as described below.

When a glass in a high-temperature state is cooled, it is thought that the glass is frozen in the structure of the high-temperature state. When a glass in this state is maintained in the transition-temperature region, the glass structure leaves the

frozen high-temperature state and gradually approaches a structure that is peculiar to the maintenance temperature. It is thought that a glass structure which corresponds to this temperature does exist and that it varies in this way so that the structure can be expressed by a fictive temperature or structural temperature τ_f. With regard to alkali ions as well, a structural temperature τ_{R_2O} is posited that corresponds to characteristic sites that can be occupied by the alkali ions at a certain temperature. On the other hand, because the temperature coefficient of the shear modulus decreases with an increase in the temperature of the glass, it appears that alkali ions in the glass are frozen at higher-temperature sites in the case of rapidly cooled glasses so that γ shows a large value. Because α does not show any great variation, the position of the minimum value, at which there is a transition from $|\alpha| > |\gamma|$ to $|\alpha| < |\gamma|$, is shifted toward lower temperatures.

5.4.3.3.   Variation of the Delay Time over Time

Depending on the glass involved, the delay time and the temperature characteristic curve of the delay time may show variation over time. This is important for determining the performance stability of delay-line glasses.

As seen in Fig. 5.35, temporal variation of the delay time at a fixed temperature can be accurately expressed as follows:

$$\log \frac{\tau_d - \tau_{d\infty}}{\tau_{d0} - \tau_{d\infty}} = -k(t - t_0) \quad . \tag{5.45}$$

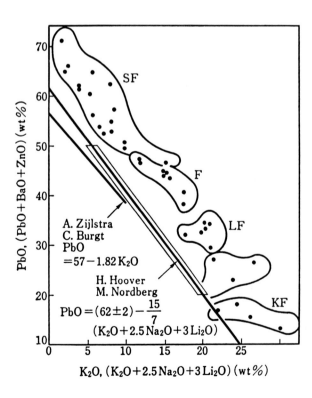

Figure 5.33.   Graph of glass compositions in which the temperature coefficient is near zero.

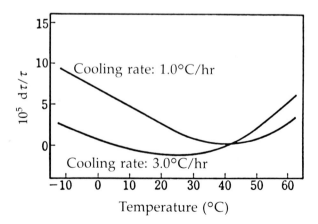

Figure 5.34.   Effect of heat treatment on temperature characteristic curve of delay time.

Here, τ_d is the delay time at an arbitrary time t, $\tau_{d\infty}$ is the delay time at saturation, and τ_{d0} is the delay time at the initial time. This equation shows a good correspondence with the following equation expressing the variation of the structural temperature:

$$\ln(\tau - T) = -Kt \quad . \tag{5.46}$$

Equation (5.49) indicates that temporal variation of the delay time occurs as a result of some type of relaxation of the structural temperature. Furthermore, the fact that temporal variation occurs even at room temperature and the fact that the activation energy is approximately 16 kcal/mole make it possible to think of this temporal variation as a stabilization process in which alkali ions frozen by rapid cooling are shifted from sites that are stable at higher temperatures to sites that are stable at lower temperatures by being maintained at room temperature. The fact that there is little temporal variation in glasses cooled sufficiently slowly also suggests that this temporal variation is a stabilization process.

5.4.3.4.   Absorption of Ultrasonic Waves

The attenuation of ultrasonic waves in a delay-line glass does not result solely from absorption by the glass, but is measured as the sum of this plus attenuation by the transducers, attenuation by the adhesive layers, and attenuation resulting from the electronics. The causes of ultrasonic absorption are (1) deformation of the glass network, (2) vibration of the alkali ions in the glass, and (3) phonon-phonon interaction. Absorption arising from cause (3) is seen in TeO_2 glasses[30]; there is no temperature dependence so that no peaks are observed in specific temperature regions. This might be called "base absorption." Absorption in delay-line glasses is thought to arise from causes (1) and (2).

Absorption Caused by Deformation Movement of the Glass Network.   Absorption caused by the deformation of the SiO_2 network has been measured by Strakna.[37] At 20 MHz, as shown in Fig. 5.36, an absorption peak was found to be present at approximately 40 K. As the frequency decreases, the peak is shifted toward lower temperatures. Accordingly, there is little effect of the network deformation on the ultrasonic absorption at room temperature, though there is a possibility that the wing of the peak might contribute slightly to ultrasonic absorption. The activation energy of electrode deformation movement is estimated to

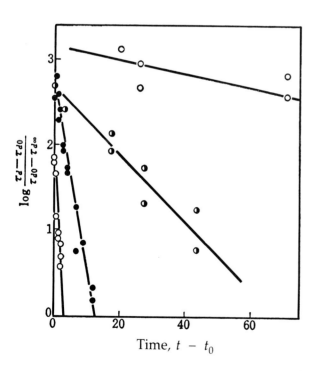

Figure 5.35.   Temperature dependence of temporal variation of delay time.

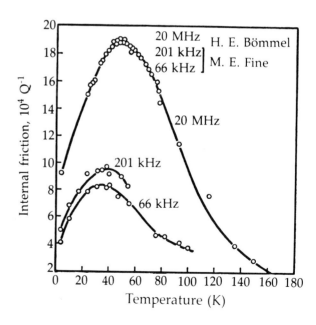

Figure 5.36.   Ultrasonic absorption by SiO_2 glass in the high-frequency region.

be approximately 0.5 kcal/mole from the relationship $f = f_0 \exp(-E/kT)$.

Absorption by Alkali Ions. When Na_2O is added to SiO_2, peaks are generated at 0 to 100°C and at 300 to 400°C, at a frequency of 1 kHz (Ref. 38). As shown in Fig. 5.37, the magnitude of the internal friction increases with the amount of alkali, and the positions of the peaks are shifted toward lower temperatures. As the frequency increases, the peaks are shifted toward higher temperatures. However, the peak that contributes most to ultrasonic absorption in delay-line glasses is the absorption peak in the vicinity of 0 to 100°C; accordingly, the effective control of this absorption is most important. In regard to this low-temperature absorption, K_2O shows less attenuation than Na_2O (see Fig. 5.38).

In regard to SiO_2–alkali systems, there are no examples of the simultaneous measurement of the SiO_2 and alkali peaks. However, the data of Kurjian[39] is available for GeO_2–alkali systems such as those shown in Fig. 5.39. The GeO_2 absorption peak is located at approximately 150 K; when Na_2O is added, the peak arising from GeO_2 decreases, and almost disappears at 10 mol% [Na_2O]. When the amount of Na_2O is increased further to 30 to 35 mol%, a large peak is generated in the vicinity of 450 K, and absorption in the vicinity of room temperature is also high. The effect seen in GeO_2–Na_2O systems probably occurs in SiO_2–Na_2O systems as well.

When alkalies are mixed, the absorption peaks seen in single-alkali glasses disappear, and absorption occurs in the temperature region of 200 to 300°C. Accordingly, absorption in the vicinity

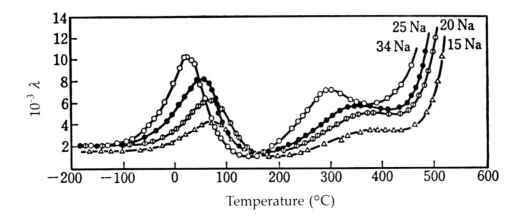

Figure 5.37. Relationship between internal friction and alkali-ion content in alkali-silicate glasses.

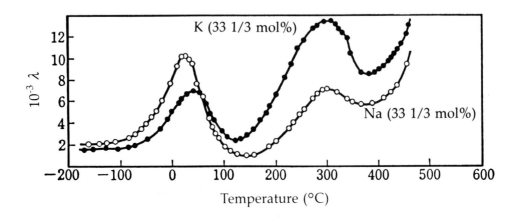

Figure 5.38. Internal friction in alkali-silicate glasses: effect of alkali-ion type.

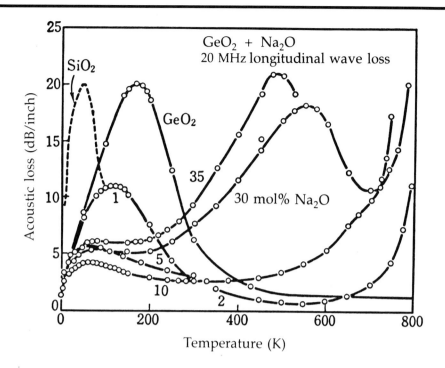

Figure 5.39.   Ultrasonic absorption in GeO₂–Na₂O systems.

of room temperature can be decreased by using mixed alkalies.[38] Furthermore, when alkali-earth ions are added to alkali silicate glasses, absorption occurs in the vicinity of 250 to 300°C at 0.4 Hz (Refs. 40, 41). As the frequency increases, the peaks move toward even higher temperatures. Accordingly, it appears that alkali-earth ions have the effect of decreasing absorption.

Figure 5.40 shows a model of the ultrasonic absorption occurring in the vicinity of 10 MHz. It appears that the factor governing absorption in the vicinity of room temperature is the wing of the alkali absorption peak, and that absorption by the network component SiO₂ and absorption by the divalent component PbO can be ignored.

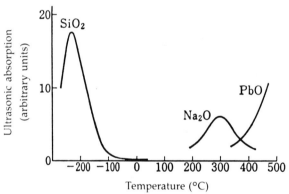

Figure 5.40.   Predicted absorption peak positions of glass constituent oxides in the vicinity of 10 MHz.

5.5.   Transparent Crystallized Glasses

5.5.1.   What are the Nuclei in the Glass Ceramics?

The invention of the glass ceramics by Stooky[42] took place in 1959. Stooky added TiO₂ to an Li₂O–Al₂O₃–SiO₂ glass as a nucleating agent, and uniform microcrystals were deposited, but no titanium oxide was found in the deposited crystals. Soon afterward, Vogel,[43] using an electron microscope, found that liquid droplets are deposited in the heat treatment process of crystallized glasses, indicating that phase separation occurs. It was hypothesized that the phase-separated liquid droplets might act as nuclei for crystallization. Izumitani et al.,[44] had doubts concerning this. The reasons for these doubts were as follows: the formation energy of nuclei in a phase transition is given by the following equation[45]:

$$\Delta G = \frac{16\pi\sigma^3}{3(\Delta G_v)^2} \quad . \tag{5.47}$$

Here, σ is the surface energy of the melt-crystal interface, and ΔG_v is the difference in free energy per unit volume between the melt and the crystal. The nucleation energy G^* in a case where crystals are formed at the interface between the melt and a foreign substance as shown in Fig. 5.41 is given by the following equations[43]:

$$\Delta G^* = \Delta G \cdot f(\theta) \quad . \tag{5.48}$$

$$f(\theta) = \frac{(2 + \cos\theta)(1 - \cos\theta)^2}{4} \quad . \tag{5.49}$$

The term θ is the contact angle; nucleation becomes easier as the contact angle decreases, i.e., as wetting becomes easier.

Let us compare a case where the foreign substance is a liquid with a case where the foreign substance is a crystal. In a case where crystals are deposited at the interface between the foreign substance and the melt, the following relationship results from the surface tension balance:

$$\cos\theta = \frac{(\sigma_{lH} - \sigma_{cH})}{\sigma_{cl}} \quad . \tag{5.50}$$

When the foreign substance is a liquid droplet, the surface energy σ_{lH} between the liquid droplet and the melt is small; accordingly, we may consider that $\sigma_{lH} < \sigma_{cH}$. In this case, $\theta > \pi/2$ [from Eq. (5.50)]. On the other hand, when the foreign substance is a crystal and the crystal-crystal surface energy is small, $\sigma_{lH} > \sigma_{cH}$. In this case, from Eq. (5.50), $\theta < \pi/2$, and $f(\theta)$ in Eq. (5.49) decreases with a decrease in θ. Accordingly, the formation energy for a heterogeneous nucleus is lower when the crystal is formed from a crystal interface than when the crystal is formed from a liquid droplet interface so that nucleation is easier in the former

case. This is the primary reason for the above mentioned doubts. Furthermore, this fact has been experimentally confirmed. It was confirmed by Doherty et al.,[46] that the nuclei in an $Li_2O-Al_2O_3-SiO_2$ system are pyrochlore $[Al_2(Ti_mZr_n)_2O_7]$. Nakagawa et al.,[44] have separated the nucleus substance from samples and have also confirmed that the crystal nuclei are pyrochlore by x-ray diffraction and electron beam diffraction.

5.5.2. Relationship between Phase Separation and Crystallization

What is the relationship between phase separation and crystallization? Must phase separation occur as a precursory phenomenon before crystallization? This problem has been the subject of much discussion by many researchers. However, this relationship has been elucidated by Nakagawa et al.[44] Phase separation and crystallization are basically unrelated, independent phenomena. As a result of phase separation, however, the composition of the melt may change so that crystals are more readily deposited. Let us explain this in the following sections.

5.5.2.1. Li_2O-SiO_2 Glasses

In the case of SiO_2-Li_2O systems (Ref. 44) (Fig. 5.42), glasses with an $Li_2O-2.5SiO_2$ composition have a region of immiscibility in the vicinity

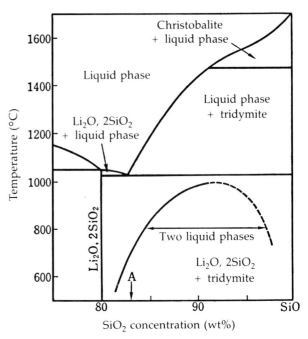

Figure 5.42. Phase diagram of SiO_2-Li_2O systems.

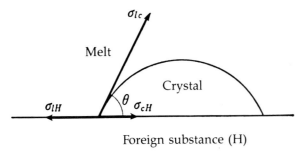

Foreign substance (H)

Figure 5.41. Heterogeneous nucleation at an interface.

of SiO$_2$ and Li$_2$O–2SiO$_2$ (Ref. 48). Accordingly, heat treatment causes such glasses to separate into liquid droplets with a high SiO$_2$ concentration and a matrix phase with a composition close to Li$_2$O–2SiO$_2$. The Li$_2$O–2SiO$_2$ crystals are deposited from the matrix phase. In such a glass, as shown in Fig. 5.43, both formation and growth of liquid-droplet nuclei, respectively, take place. There is no difference between the number of Li$_2$O–2SiO$_2$ crystals deposited from a rapidly cooled glass that contains no liquid droplets and the number of Li$_2$O-2SiO$_2$ crystals deposited from a glass that has been heat treated so that liquid droplets were formed. It is seen from this that a liquid-liquid interface does not promote heterogeneous nucleation. In other words, there is no direct formation of nuclei from the surfaces of liquid droplets.

5.5.2.2. Li$_2$O–SiO$_2$–TiO$_2$ Glasses

When a glass with a composition formed by adding 22.5% TiO$_2$ to Li$_2$O · 2.5SiO$_2$ is heat treated at 600°C, crystallization progresses into the matrix phase from the liquid-droplet interfaces created by liquid-liquid separation (see Fig. 5.44) (Ref. 44). The crystals contained in a sample heat-treated for 30 minutes at 600°C are Li$_2$O · TiO$_2$ and range in size from 20 to 50 Å. In a sample heat treated for two hours at 600°C, large amounts of Li$_2$O · 2SiO$_2$ crystals are deposited in the matrix phase in addition to small amounts of Li$_2$O · TiO$_2$ crystals. In this glass, Li$_2$O · TiO$_2$ crystals are deposited only in the temperature range extending from 550 to 715°C, whereas Li$_2$O · 2SiO$_2$ crystals are deposited in the temperature region extending fron 600 to 900°C. If the glass is heat treated at 750°C, where Li$_2$O · TiO$_2$ crystal deposition does not occur, only a very small number of Li$_2$O · 2SiO$_2$ crystals are deposited. Accordingly, in

this glass as well, heterogeneous nucleation of Li$_2$O · 2SiO$_2$ crystals does not occur at liquid-liquid interfaces; in this case, the Li$_2$O · TiO$_2$ crystals formed in the matrix phase near the liquid-droplet interfaces act as crystal nuclei. Because Li$_2$O · TiO$_2$ crystals are not formed even if this glass is slowly cooled from the melting temperature, it appears that the initial formation of Li$_2$O · TiO$_2$ crystals that act as crystal nuclei results from liquid-liquid separation.

5.5.2.3. Li$_2$O–Al$_2$O$_3$–SiO$_2$–(TiO$_2$, ZrO$_2$) Glasses

The relationship between liquid-liquid separation and crystallization was investigated for a glass in which TiO$_2$ and ZrO$_2$ (2% each) were added as crystal nucleating agents to a basic composition of 75% SiO$_2$, 4.5% Li$_2$O, and 20.5% Al$_2$O$_3$. This glass crystallizes according to the process

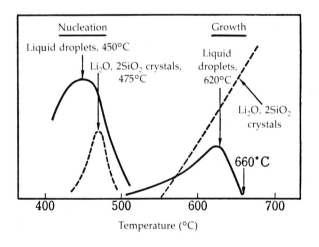

Figure 5.43. Liquid-liquid separation and crystallization in a glass with an Li$_2$O · 2.5SiO$_2$ composition.

(a) Rapid cooling   (b) 30 min at 600°C   (c) 2 hr at 600°C

Figure 5.44.   Electron micrographs of glass samples (composition obtained by adding 22.5% TiO$_2$ to Li$_2$O · 2.5SiO$_2$) heat treated at 600°C (Ref. 42).

shown in Fig. 5.45. Heat treatment causes the glass to undergo liquid-liquid phase separation; the liquid droplet formation rate reaches a maximum at 715°C, whereas the liquid droplet growth rate reaches a maximum at 765°C. The miscibility temperature of this glass is 800°C; no liquid droplets are present at temperatures higher than this. The principal crystal deposited from this glass is a β-quartz solid solution; this crystal grows at temperatures exceeding 780°C. Furthermore, in the temperature region extending from 740 to 800°C, a different type of crystal from the β-quartz solid solution is formed. As in the previous description, this different crystal is pyrochlore [$Al_2(Ti_mZr_n)_2O_7$]. In this case, the size of the individual pyrochlore crystals ranges from 25 to 30 Å (as determined by small-angle x-ray scattering). Because these pyrochlore crystals are not formed at temperatures above the miscibility temperature of the glass, it appears that the formation of pyrochlore crystals that act as crystal nuclei is made possible by liquid-liquid separation. Accordingly, it appears that in this glass as well, the liquid-liquid interface is not the starting point of crystal nucleation; in this case, the pyrochlore type microcrystals deposited in the matrix phase act as crystal nuclei, and the deposition of these pyrochlore type microcrystals is made possible by liquid-liquid separation of the glass.

The deposition of crystal nucleus $Li_2O \cdot TiO_2$ or pyrochlore type crystals accompanying liquid-liquid separation is apparently attributable to the following mechanism occurring in the vicinity of interfaces between the liquid droplets and the matrix. A diffusion layer with an SiO_2 concentration lower than that of the matrix phase is present as a

result of SiO_2 deposition[49]; accordingly, the concentrations of Li_2O, TiO_2, ZrO_2, and Al_2O_3 are increased.

5.5.3. Formation Mechanism of SiO_2–Li_2O–Al_2O_3 Crystallized Glasses

The principal crystal contained in SiO_2–Li_2O–Al_2O_3 crystallized glasses is a β-quartz solid solution; this is a hexagonal crystal with a structure in which a portion of the Si^{4+} of the β-quartz is replaced by Li^+ and Al^{3+}. The crystallization process appears to proceed as follows in a glass formed by adding TiO_2 and ZrO_2 (2% each) to a basic glass composition of 75% SiO_2, 4.5% Li_2O, and 20.5% Al_2O_3 (Ref. 50). As shown in Fig. 5.46, liquid droplets with a high SiO_2 concentration are formed by liquid-liquid separation in the early stages of heat treatment. Next, pyrochlore type microcrystals [$Al_2(Ti_mZr_n)_2O_7$] are deposited in the matrix phase. With these crystals acting as nuclei, β-quartz solid-solution crystals are deposited in the matrix phase. The concentration of SiO_2 in the β-quartz solid-solution crystals increases as the heat treatment progresses; the reason for this is that the liquid droplets generated by liquid-liquid separation are gradually transformed into β-quartz solid-solution crystals. When a glass of this type is heat treated at a temperature above the miscibility temperature, the liquid droplets with a high SiO_2 concentration are all transformed into β-quartz solid-solution crystals so that no liquid droplets remain in the crystallized glass.

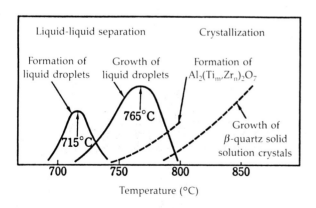

Figure 5.45. Liquid-liquid separation and crystallization in an SiO_2–Li_2O–Al_2O_3 glass to which TiO_2 and ZrO_2 have been added.

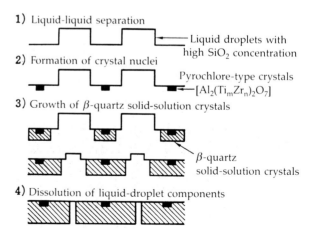

Figure 5.46. Formation process of SiO_2–Li_2O–Al_2O_3 crystallized glass with added TiO_2 and ZrO_2.

Figure 5.47. Electron micrograph of SiO$_2$–Li$_2$O–Al$_2$O$_3$ crystallized glass containing TiO$_2$ and ZrO$_2$.

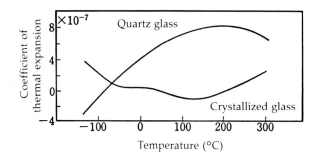

Figure 5.48. Coefficients of thermal expansion of quartz glass and SiO$_2$–Li$_2$O–Al$_2$O$_3$ crystallized glass.

5.5.4. Properties of Transparent Crystallized Glasses

An SiO$_2$–Li$_2$O–Al$_2$O$_3$ crystallized glass produced in this way consists of about 80% β-quartz solid solution and 20% vitreous material. The β-quartz solid solution crystals range in size from 20 to 40 nm so that this crystallized glass is transparent. Figure 5.47 shows an electron micrograph of such a transparent crystallized glass, and Fig. 5.48 shows the coefficient of thermal expansion of this glass compared with that of quartz glass. The reason that the expansion coefficient is close to zero is that the β-quartz solid solution has a negative expansion coefficient, whereas the remaining matrix glass has a positive expansion coefficient so that the two values cancel each other. Examples of transparent crystallized glasses that have actually been put into production are Crystron-O (Hoya) and Zerodur (Schott).

5.6. Postscript: Other New Glasses

In addition to the glasses described above, I had also originally intended to write about the following new glasses:

1. Photosensitive glasses
2. Optical fibers
3. GRIN lenses
4. Photochromic glasses
5. Pressed lenses
6. Filter glasses

Also of interest to me were recently developed glasses that include pressed lenses, hydrated glasses, fluoride glasses, nitrogen-containing glasses, metallic glasses, glasses for use *in vivo*, and sol-gel reaction glasses, etc. I had hoped to write about these glasses as well, but time did not permit. A more suitable person will doubtless master these glasses as time passes and as the evaluation of the new glasses is completed. As for my obligation as a single researcher engaged in the study of optical glasses, I must say that this is enough.

"Unless a grain of wheat drops into the earth and dies, it remains single, but if it dies, it produces a rich yield" (John 12:24, The New Testament). I hope from the bottom of my heart that a rich yield of fruit will be born and will grow from this little seed of mine.

I had hoped to complete this work while my mother was still alive; I deeply regret that I was unable to do so.

On the plane from
Anchorage, Alaska
(November 11, 1983).

References

1. Japanese Patent Kokai No. 53-144913.
2. Japanese Patent No. 53-42330.
3. Japanese Patent Nos. 54-6042 and 54-6043.
4. Japanese Patent No. 47-16811.
5. Japanese Patent Nos. 50-8445, 50-8446, and 50-8448.
6. Japanese Patent No. 53-28169.
7. T. Izumitani and K. Nakagawa, *VIIth International Glass Congress*, Brussels, Belgium, 1965 (Gordon and Breach, NY), 5-1.
8. S. Hirota and T. Izumitani, *Yogyo-Kyokai-shi* **84**, 435 (1976).
9. S. Hirota and T. Izumitani, *J. Non-Cryst. Sol.* **29** (1), 109 (1978).
10. A. J. Bourdillon, F. Khumalo, and J. Bordas, *Phil. Mag. B* **37**, 731 (1978).
11. I. H. Malitson, *J. Opt. Soc. Am.* **55**, 1205 (1965).
12. T. Izumitani and H. Toratani, *J. Non-Cryst. Sol.* **40** (1–3), 611 (1980).
13. H. Toritani and T. Izumitani, *Kogaku (Optical Science)* **8** (6), 372 (1979).
14. W. F. Krupke, *IEEE J. Quantum Electron* **OE 10**, 450 (1974).
15. T. Izumitani and Tsuru, *Reza Kenkyu (Laser Research)* **2** (3), 168 (1971).
16. S. Stokowskii, *Laser Program Annual Report*, Lawrence Livermore National Laboratory, Livermore, CA, UCRL-50021-79, Vol. **1**, 2-153 (1979).
17. W. J. Weber, *J. Non-Cryst. Sol.* **47** (1–2), 117 (1982).
18. A. G. Aranesov, T. T. Basiev, Yu. K. Voron'ko, B. I. Denker, A. Ya. Karasik, G. V. Maskimova, V. V. Osiko, V. F. Pisarenko, and A. M. Prokhorov, *Sov. Phys., JETP* **50** (5), 886 (1979).
19. R. D Peacock, *Structure and Bonding 22* (Springer-Verlag, Berlin, Heidelberg, New York, 1975).
20. C. K. Jorgensen, *Modern Aspect of Ligand Field Theory* (North-Holland, Amsterdam, 1971).
21. T. Izumitani, H. Toratani, and H. Kuroda, *J. Non-Cryst. Sol.* **47** (1), 87 (1982).
22. C. B. Layne, W. H. Lowdermilk, and M. J. Weber, *Phys. Rev.* **B16**, 10 (1977).
23. H. Bequerel, *Compt. Rend.* **125**, 679 (1897).
24. J. H. van Vleck and M. H. Hebb, *Phys. Rev.* **46**, 17 (1934).
25. M. J. Minot, *J. Opt. Soc. Am.* **66**, 515 (1976).
26. Y. Asahara and T. Izumitani, *J. Non-Cryst. Sol.* **42** (1–3), 269 (1980).
27. S. P. Mukherjee and W. H. Lowdermilk, *CLEO '81 WP-2 June 10–12, Washington D.C.* (1981).
28. L. M. Cook, W. H. Lowdermilk, D. Milam, and J. E. Swain, *CLEO '82 WM-1 April 14–16, Washington D.C.* (1982).
29. R. Keck, W. Seka, L. M. Goldman, J. M. Sourec, R. Boni, L. Forsley, R. S. Craxton, J. A. Delettrez, and R. L. MacCrory, *CLEO '81 WH-2 June 11–13, Washington D.C.* (1981).
30. Watanabe and T. Izumitani, *Oyo Butsuri (Applied Physics)* **41**, 402 (1972).
31. I. Masuda and T. Izumitani, *Xth International Glass Congress, Kyoto, Japan* 5-74 (1974).
32. T. O. Woodruff and H. Ehrenreich, *Phys. Rev.* **123**, 1553 (1961).
33. S. Spinner, *J. Am. Ceram. Soc.* **45** (8), 394 (1962).
34. E. van Deeg, *Glastech. Ber.* **31**, 229 (1962).
35. H. L. Hoover and M. E. Nordberg, U.S. Pat., No. 3154425.
36. A. L. Zijlstra and C. M. van der Burgt, *Ultrasonics* **29**, (1967).
37. R. E. Strakna and H. T. Savage, *J. Appl. Phys.* **35** (5), 1445 (1964).
38. O. L. Anserson and H. E. Bömmel, *J. Am. Ceram. Soc.* **38** (4), 125 (1955).
39. C. R. Kurjian and J. T. Krause, *J. Am. Ceram. Soc.* **49**, 134 (1966).
40. R. J. Ryder and G. E. Rindone, *J. Am. Ceram. Soc.* **43** (12), 662 (1960).
41. R. J. Ryder and G. E. Rindone, *J. Am. Ceram. Soc.* **44** (11), 532 (1961).
42. S. D. Stooky, *Glastech. Ber.* **32K** VI (1959).
43. W. Vogel and K. Gerth, *Symposium on Nucleation and Crystallization in Glasses and Melts, American Ceramic Society, Inc.* (1964), p. 11.
44. K. Nakagawa and T. Izumitani, *Phys. Chem. Glass.* **10**, 179 (1969).
45. D. Turnbull and M. H. Cohen, *Modern Aspects of the Vitreous State I* (1960), p. 38.
46. P. E. Doherty, D. W. Lee, and R. S. Davis, *J. Am. Ceram. Soc.* **50** (2), 77 (1967).

47. Horikawa and K. Nakagawa, *Dai-6-kai Yogyo Kiso Toronkai Yoshishu (6th Glass Fundamentals Symposium Summaries)* (1968), p. 13.
48. Y. Moriya, D. H. Warington, and P. W. Douglas, *Phys. Chem. Glass.* **8**, 20 (1967).
49. J. J. Hummel and S. M. Ohlberg, *J. Appl. Phys.* **36**, 144 (1965).
50. K. Nakagawa and T. Izumitani, *J. Non-Cryst. Sol.* **7**, 168 (1972).

APPENDIX A
Reflection Spectra of Glass in the Vacuum and Extreme Ultraviolet

The relation between absorption and dispersion of glass is ordinarily expressed as[2]

$$n^2 - 1 = \frac{KN_1 f_1}{1/\lambda_1^2 - 1/\lambda^2} + \frac{KN_2 f_2}{1/\lambda_2^2 - 1/\lambda^2}$$
$$+ \frac{KN_3 f_3}{1/\lambda_3^2 - 1/\lambda^2} + \frac{KN_4 f_4}{1/\lambda_4^2 - 1/\lambda^2} ,$$

where $K = e^2/\pi m c^2$, N is the number of relevant ions per unit volume, f is the oscillator strength, and λ is the wavelength. Subscripts 1 and 2 refer to absorption in the vacuum ultraviolet (VUV) by bridging and nonbridging oxygen, respectively, subscript 3 to absorption by highly polarizable cations (such as Pb^+), and subscript 4 to absorption in the infrared by interatomic vibration.

To express the dispersion precisely, however, it is necessary to include a fifth term (with subscript zero) representing absorption in the extreme ultraviolet (EUV, beyond 10 eV) by bridging ions, so that the expression becomes

$$n^2 - 1 = \frac{KN_0 f_0}{1/\lambda_0^2 - 1/\lambda^2} + \frac{KN_1 f_1}{1/\lambda_1^2 - 1/\lambda^2}$$
$$+ \frac{KN_2 f_2}{1/\lambda_2^2 - 1/\lambda^2} + \frac{KN_3 f_3}{1/\lambda_3^2 - 1/\lambda^2} + \frac{KN_4 f_4}{1/\lambda_4^2 - 1/\lambda^2} .$$

For fused SiO_2, λ_0 is found at about 17.8 eV (68 nm). In silicate glasses, λ_1 is found at about 10.2 eV (122 nm). Depending on the modifier ions,[3] λ_2 is found between 6.9 and 8.7 eV (179 to 142 nm). In lead silicate glasses, λ_3 is found at 5.96 eV (208 nm), and in silicate glasses, λ_4 is found at 0.13 eV (9.8 μm).

The reflection spectrum of fused SiO_2, which has been measured by Philipp,[4] Sigel,[5] and Bourdillon et al.,[6] is shown in Fig. 1. There are strong reflection bands at 17.8 and 10.4 eV. Where are the corresponding bands in silicate, phosphate, borate, aluminum fluoride, and zirconium fluoride glasses? It is well known that the Abbe number ν (the inverse of the dispersion) increases in the order silicate < borate < phosphate glasses, so it is anticipated that UV absorption wavelengths will increase in the order phosphate < borate < silicate. It is also anticipated that the EUV absorptions of fluoride glasses should fall at shorter wavelengths than those of oxide glasses, but this cannot be confirmed from reflection spectra below 10 eV. This is why we measured EUV reflection spectra.

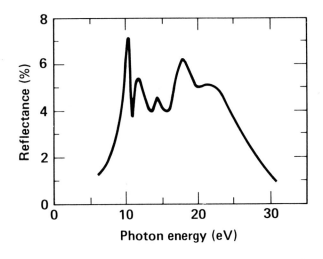

Figure 1. EUV reflection spectrum of fused SiO_2.

Measurement of Reflection Spectra by Synchrotron Radiation

To measure reflection spectra in the EUV we used synchrotron radiation (SR). SR is the electromagnetic wave radiated in the tangential direction when high-speed electrons are deflected by magnetic or electric fields in ultrahigh vacuum. SR is a strong, continuous light source in the EUV and soft x-ray regions.

Figure 2 shows the SR vacuum spectrometer used in these measurements. Incident radiation is dispersed by the toroidal grating monochromator. The monochromatic light is reflected from the sample with an angle of incidence of 14°, and is detected by a sodium salicylate plate and a photomultiplier. The apparatus was evacuated to 10^{-8} Torr. Figure 3 shows the incident SR spectrum.

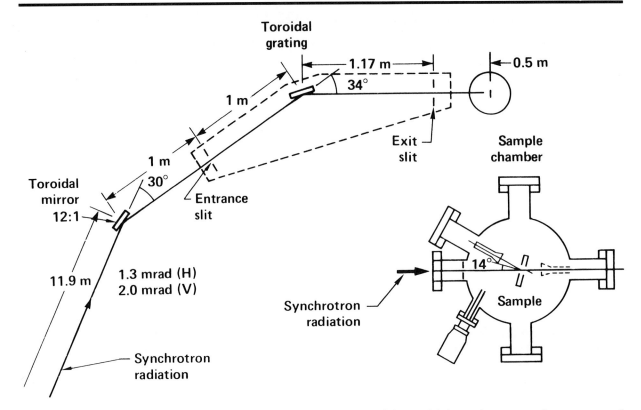

Figure 2.   Schematic diagram of the vacuum spectrometer with toroidal-grating monochromator and sample chamber.

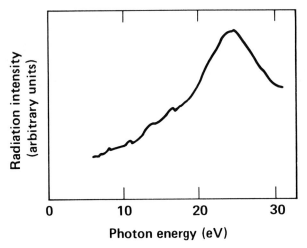

Figure 3.   Incident SR light spectrum between 10 and 35 eV.

Reflection Spectra of Fused SiO$_2$, Aluminum Fluoride, and Zirconium Fluoride Glasses

Figure 4 shows the reflection spectra of fused SiO$_2$, aluminum fluoride, and zirconium fluoride glasses. Table 1 lists the locations of the reflection bands in these glasses and gives the Abbe and n_d

Table 1.   Reflection bands and Abbe and n_d values in fused SiO_2, aluminum fluoride, and zirconium fluoride glasses.

	Reflection bands (eV)				Abbe number v	n_d
Fused SiO_2 glass	10.3	11.6	14.3	17.8	67.8	1.4585
AlF_3–CaF_2 glass	11.4	14.5	17.0	23.0	90.8	1.4565
ZrF_4–BaF_2 glass	11.6	13.2	19.2	23.2	75.3	1.5210

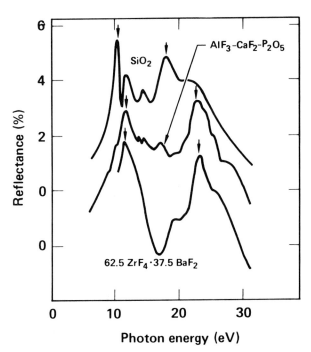

Figure 4.   EUV reflection spectra of fused SiO_2, aluminum fluoride, and zirconium fluoride glasses.

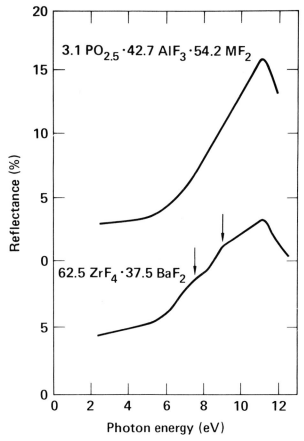

Figure 5.   VUV reflection spectra of aluminum fluoride and zirconium fluoride glasses.

values. The intensity of reflection bands is strong at 10.3 and 17.8 eV in fused SiO_2, while the reflectivity is strong at λ_1 = 11.4 eV and λ_0 = 23 eV in the two fluoride glasses. The higher photon energy at λ_1 and the higher value of KN_0f_0 at λ_0 give the fluoride glasses higher Abbe numbers than that of fused SiO_2, as shown in Table 1. The difference in the Abbe numbers of AlF_3–CaF_2 and ZrF_4–BaF_2 glasses is due to the fact that AlF_3–CaF_2 glass does not have absorption bands in the VUV below 10 eV, as shown in Fig. 5, while ZrF_4–BaF_2 glass has reflection bands at 9.3 and 7.4 eV, which give it higher dispersion.

Reflection Spectra of Barium Silicate, Borate, and Phosphate Glasses

Figures 6 and 7 show the reflection spectra of barium silicate, borate, and phosphate glasses with the same composition in the ranges 2 to 31 eV and 2 to 12 eV, respectively. Table 2 lists the locations of the reflection bands in these glasses, and gives the Abbe and n_d values.

The reflection spectra of the borate and phosphate glasses in the EUV are very similar, but the bridging-oxygen reflection band λ_1 is at a lower energy in the borate glass (9.0 eV) than in the phosphate glass (11.0 eV). Accordingly, the Abbe number of the borate glass is lower than that of the phosphate glass.

It seems anomalous that the reflection band λ_1 is at a little higher energy in silicate glass than in phosphate glass, while the Abbe number is

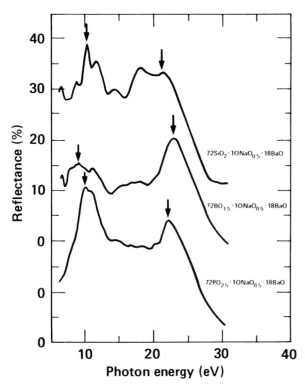

Figure 6.   EUV reflection spectra of barium silicate, borate, and phosphate glasses.

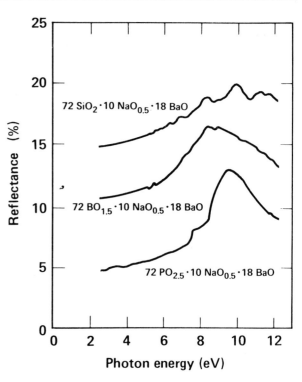

Figure 7.   VUV reflection spectra of barium silicate, borate, and phosphate glasses.

Table 2.   Reflection bands and Abbe and n_d values in barium silicate, borate, and phosphate glasses with the same composition.

	Reflection bands (eV)				Abbe number v	n_d
Silicate glass	10.3	11.7	18.0	21.2	58.19	1.55687
Borate glass	9.0	11.2	18.0	22.8	60.41	1.61203
Phosphate glass	10.1	11.0	18.1	22.0	65.50	1.54623

smaller in silicate than in phosphate glass. However, as shown in Fig. 7, the intensity of the reflection band λ_2 due to nonbridging oxygen is higher in silicate than in borate or phosphate glasses, so that the contribution of the absorption bands λ_2 to dispersion is stronger in silicate glass.

Thus the fact that the Abbe number of optical glass is higher in the order phosphate > borate > silicate glasses is now understood on the basis of the measured VUV and EUV reflection spectra of these glasses.

References

1. T. Izumitani, S. Hirota, and H. Onuki, submitted to *J. Noncryst. Solids*.
2. T. Izumitani, *Optical Glass*, Lawrence Livermore National Laboratory, Livermore, Calif., UCRL-TRANS-12065 (1965), p. 2-12, Eq. (2.12).
3. T. Izumitani and S. Hirota, *IInd Internationales Otto-Schott-Kolloquium*, Jena, DDR (1982).
4. H. R. Philipp, *Solid State Commun.* **4**, 73–75 (1966).
5. G. H. Sigel, Jr., *J. Phys. Chem. Solids* **32**, 2373–2383 (1971).
6. A. J. Bourdillon, F. Kumalo, and J. Bordas, *Phil. Mag B* **37**, 731–738 (1978).

INDEX